Praise for *Body of Work* by Christine Montross

"To the fine essays of the poet and undertaker Thomas Lynch must be added this gleaming, humane work by the poet Christine Montross, written during, and about, her first year of medical school. . . . The chilly detachment of medical instruction . . . makes Montross's writing shine brighter. . . . The author dissects her own emotions as deftly as she does the organs and ligaments of the cadaver, her pen as revelatory as her scalpel. . . . Medical professionals will find much to comfort, but also to challenge themselves in these pages. The book is of even more value to patients. . . . Eloquent and persuasive."
—Mary Roach, *The New York Times Book Review*

"In this unflinching record of her first semester at Brown's med school, [Montross] recalls battling nightmares during the months she dissected an elderly woman's remains. . . . She supplements her graphic lab observations with insightful reflection and research. . . . Throughout, her writing is sparklingly lucid. A–"   —*Entertainment Weekly*

"This is a new voice in medical writing: lyrical, insightful, introspective. Montross, by probing deeply into the hidden recesses of the body, brilliantly illuminates the soul. A welcome debut."
—Jerome Groopman, MD, Recanati Professor of Medicine, Harvard University and author of *How Doctors Think*

"Montross was a poet before she was a doctor, and her language in *Body of Work*, an exceptionally thoughtful memoir about the first semester of medical school, is as precise as her scalpel cuts become by the final exam. . . . We should be grateful, too—especially those of us who squirm away from the physical truths of our existence—for this beautiful book and the glimpse it offers of a place off limits to anyone without Montross's clearsighted courage."   —*The Washington Post*

"The table and the blade, blood and bodies, dissection and discernment—such are the properties of the medical arts. From her hands-in, hands-on study of parts, whole persons emerge in Dr. Montross's wonderfully curious text. Here are ample doses of metaphor and good medicine."

> —Thomas Lynch, author of *Booking Passage*, *The Undertaking*, and *Bodies in Motion and at Rest*

"Dr. Montross, a recent medical graduate, has tethered an earnest and readable reflection on the process of becoming a doctor to the methodical dissection of a human cadaver, the first of all too many professional initation rites. . . . Who better to describe the strange terrain between doctor and patient, dissector and cadaver, us and them, than someone actually crossing that no-man's-land?"

> —*The New York Times*

"The physician, like the sculptor, approaches the human body with reverence and admiration. Carried a little further, it becomes worship. In *Body of Work*, an unflinching memoirist conveys the process, both emotional and intellectual, by which human anatomy is mastered by the doctor-to-be. It should be read by anyone with aspirations for a life in medicine."

> —Richard Selzer, author of *Mortal Lessons*, *The Doctor Stories*, and *Letters to a Young Doctor*

"In her well-wrought first book, *Body of Work*, Christine Montross—a psychiatry resident at Brown University—has produced an unusual synthesis of these types of narrative. Part memoir, part love letter, part medical history, part rationalization and part poetry, *Body of Work* resists the silences of the medical profession to explore the author's relationship to the cadaver she dissected, one to whom she gives the loaded name of Eve. . . . *Body of Work* is at its best when Montross, who is also a poet, allows us to observe the astonishing beauty her dissection reveals, and to relish the language she uses to describe it. . . . Equally gripping are the stories she shares of loved ones who fall ill, and later, of some of her first patients, whose living-and-breathing bodies insistently remind her of Eve's—and her own—humanity. Here, language, arising from the body, becomes healing, accommodating both knowledge and wonder, abetting not only the joy of discovery but also the empathic connection between teacher and student that underlies learning anatomy, and learning to heal more generally. . . . Thought-provoking."   —*Los Angeles Times*

"How lucky we are that a poet decided to become a physician. Although all physicians share a personal history of countless hours in the human anatomy lab, only a rare few, I suspect, would be able to so deftly illuminate this transforming and peculiar experience. Montross is a master of detail, so much so that I was shocked to find myself hovering over my own cadaver in medical school again, holding a scalpel as if for the first time."

—Katrina Firlik, MD, neurosurgeon and author of
Another Day in the Frontal Lobe

"Before Christine Montross decided to become a psychiatrist, she was a poet, university writing instructor, and high school English teacher. So she has a way with words. . . . Montross brought that talent to one of the most traumatic parts of medical training—anatomy, or the dissection of the human body. . . . Mixed into Montross' ruminations on the human body and the nature of life is a fascinating study of the history of the dissection of bodies by medical students."—The Associated Press

"This is a book about crossing the bar. The anatomies discussed here are diverse and gripping, and remind me of the essays of Richard Selzer, which is a high compliment indeed."

—Edward Hoagland, author of *Compass Points*

"Ever thoughtful in her feeling, ever feeling in her thinking, Montross communicates the hopes and the limits of medicine, the strivings and pitfalls of doctoring. Honest and informative, Montross' first book ranks among the best of medical memoirs."   —*Portsmouth Herald*

"I never expected to feel nostalgia about gross anatomy lab, but Christine Montross' thoughtful book . . . evokes just that sense. Montross . . . chronicles the dissection of her lab group's cadaver (Eve) with unflinching honesty. . . . The vivid descriptions of the dissection itself will interest nonphysicians, providing an educational glimpse in much the way that the popular touring exhibition Bodies does. For physicians, the book resonates on a more emotional level. Montross' lovely language, interspersed with snippets of poetry, allows us to process a powerful shared experience that few of us had the time or maturity to process as it was happening. . . . Even all these years later, I approached this book with something of the same trepidation I felt in anatomy lab—but I was delighted to find it beautiful, interesting and ultimately uplifting."
—*Diversion*

"Montross' memoir is a philosophical view of life, death and the doctor-patient relationship. . . . [T]his memoir offers a distinct view for prospective medical students on what awaits them in the human anatomy lab."
—*The Birmingham News*

"Tantalizing . . . Heart-wrenching . . . [Montross's] thoughtful meditations on balancing clinical detachment and emotional engagement will easily find a spot on the shortlist of great med school literature."
—*Publishers Weekly* (starred review)

"Very poetic."
—*The Baltimore Sun*

"[Montross] reports in vivid, often poetic detail the physical, mental and emotional demands of meticulously taking apart the dead body of a woman she calls Eve, an experience that enthralls her, exhausts her, gives her haunting dreams and teaches her human anatomy as no textbook could. . . . Not for the squeamish, but an eye-opener for would-be doctors."
—*Kirkus Reviews*

"An emotional roller coaster . . . Touching."
—*USA Today*

PENGUIN BOOKS

BODY OF WORK

Dr. Christine Montross is a resident in psychiatry at Brown University. She received her master's of fine arts in poetry from the University of Michigan and has had several poems published in literary journals. While compiling this book, she traveled to anatomical theaters, sought out holy relics, and dissected three arms, a leg, and an entire human body. She lives in Rhode Island with her partner, Deborah, and their one-year-old daughter, Maude.

CVM CAESAREAE

M.diest. Galliarum Regis, & Senatus Veneti gra-
tiæ & priuilegio, ut in diplomatis eorundem continetur.

Body of Work

Meditations on Mortality

from the Human Anatomy Lab

CHRISTINE MONTROSS

PENGUIN BOOKS

PENGUIN BOOKS

Published by the Penguin Group

Penguin Group (USA) Inc., 375 Hudson Street, New York, New York 10014, U.S.A. •
Penguin Group (Canada), 90 Eglinton Avenue East, Suite 700, Toronto, Ontario, Canada
M4P 2Y3 (a division of Pearson Penguin Canada Inc.) • Penguin Books Ltd, 80 Strand, London WC2R 0RL,
England • Penguin Ireland, 25 St Stephen's Green, Dublin 2, Ireland (a division of Penguin Books Ltd) •
Penguin Group (Australia), 250 Camberwell Road, Camberwell, Victoria 3124, Australia (a division of
Pearson Australia Group Pty Ltd) • Penguin Books India Pvt Ltd, 11 Community Centre, Panchsheel Park,
New Delhi – 110 017, India • Penguin Group (NZ), 67 Apollo Drive, Rosedale, North Shore 0632,
New Zealand (a division of Pearson New Zealand Ltd) • Penguin Books (South Africa) (Pty) Ltd,
24 Sturdee Avenue, Rosebank, Johannesburg 2196, South Africa

Penguin Books Ltd, Registered Offices:
80 Strand, London WC2R 0RL, England

First published in the United States of America by The Penguin Press,
a member of Penguin Group (USA) Inc. 2007
Published in Penguin Books 2008

1   3   5   7   9   10   8   6   4   2

For Deborah,
and for Eve

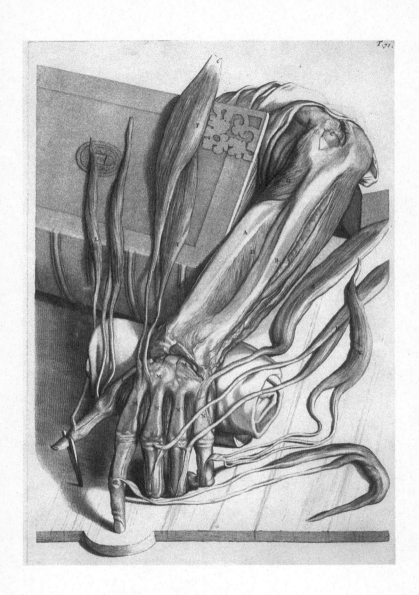

Dispel from your mind the thought that an understanding of the human body in every aspect of its structure can be given in words . . .

LEONARDO DA VINCI

Contents

Body of Work

Mystery

I have said that the soul is not
more than the body
And I have said that the body is not
more than the soul . . .

WALT WHITMAN, *LEAVES OF GRASS*

F lat-calm summer evenings on the northern Michigan lake of my childhood, I'd tug on my swimsuit and wade out in the clear green water to float. No matter how far out I walked, I could hear my family's voices on our dock, my father's deep-toned stories, my grandmother's broad, cackling laugh. But when I'd lie back in the water, arms and legs spread out like a snow angel, the lake would cover my ears with a storm of silence. I'd lie there and breathe, loving the way my body would rise and fall. I thought that I was the only moving thing in the stillness of the deep waters. I'd exhale slowly and let my body sink—my feet and legs first, then hips and chest, and just when all of me was submerged except my mouth and nose, I'd breathe in again and float back to the surface, as if gravity were a law you could choose to disobey.

What was this breath force within me? I never knew. Sometimes I would feel the water seep up to the corners of my nose and mouth; I'd keep exhaling and let my body sink to the bottom. There I'd lie, staring up at the surface, with its mercurial shine. In those moments, with my world tinged in foggy hues of green, I felt like the lone inhabitant of a sacred space: my whole young self held close and rocked by the water. But before long, out of air, I'd have to thrust myself back to the surface to float some more. When I got cold or tired, I'd walk back in, toward the voices and the laughter, wondering what domain my small breath had over water, how both air and lack of it could make me rise.

I still think of that quiet, of that sense of something powerful and unseen in me that I both could and couldn't control, of the understanding that sometimes life's truths seem to contradict each other. Now I am a student of medicine, a field with its own great paradoxes. The first of these I encountered in my anatomy class, and it is still one of the most powerful: that you begin to learn to heal the living by dismantling the dead.

The dead body harbors the great mysteries of creation and humanity: the hidden beauty and intricacy of function, the insistence of individuality, the inevitability of decline, the incontrovertibility of death set up against the ill-defined boundaries of life. Opening the body begins to unveil these mysteries. Centuries upon centuries of doctors have done so in search of wonder and knowledge.

The moment I raise a scalpel to a body is a rite of initiation. With my first cut, I have begun a personal transformation that differentiates me from my friends and family. This book is about that transformation. It is about performing previously unthinkable actions in order to discover wondrous and previously unimaginable realms. It is about joining a history of anatomy that includes grave robbers and executioners, murderers and mutants, courageous blasphemers and the bodies of saints. This book is about dissecting a dead body in the hopes of one day making living bodies more whole.

I am now in my final year of medical school, mere months away from being called "Doctor." Before I began my medical training, I was at various times a poet, a university writing instructor, a high-school English teacher to a group of troubled kids. I am older than all but a handful of my classmates and share a comparatively adult life in a little bungalow with my partner of more than seven years and our sweet old dog. I have family members who are sick, perhaps dying. I bring each of these perspectives to medicine, and indeed to the dissection table; they have informed my experience of becoming a doctor, and they have shaped what I have written here. But in the end this book comes from my interactions with the bodies of strangers, both dead and living, and the privileged view that I was given into their innermost workings and failings.

As a medical student, I have watched experts with decades of training deftly remove a superfluous length of vein from a patient's leg and transform it into a critical channel of blood for the heart. Though I was duly impressed by the surgeons' expertise, what astonished me most was the regular and insistent kicking of the patient's heart beneath the operating fingers, even when its muscular flesh was cut, bled, and sewn. The early anatomical puzzles have long ago been solved, uprooting beliefs in air coursing through veins, tiny workers stoking internal furnaces, and the hysterical uterus wandering through the body. Yet there is no shortage of mystery in the body for me today, even after having cut it apart and held its wildly various shapes and tissues in my hands.

I have been surprised by the rush of feeling that has arisen in me at the most primal of moments: the thrill of using my own much-practiced sutures and knots to close surgical wounds, the giddy exhilaration that swept over me each time my hands delivered a baby from its mother's womb, the unspeakable ache of sitting beside a man deep in the throes of dying. I have learned that the body and mind are not as easily separable as I had once imagined and that the treatment of one nearly always demands an understanding of the other. Time and again during the course of medical school, I was reminded of this, by the cardiac patient suffering from depression following his bypass surgery, by the elderly woman in the emergency room whose panic attacks deprived her of oxygen, by a brain-damaged man who no longer recognized his own left arm, by a woman who complained of chronic menstrual cramps despite the fact that her uterus had been removed years ago.

The human body harbors mysteries that are not solved by textbooks or studying, and, as I have been confronted with them, I have found myself amazed, humbled, and unnerved. My medical training thus far has put me in positions of both omnipotence and powerlessness, has revealed stark clarity and confounding darkness, has made me a vehicle of hope one day and of despair the next. I believe these intersections and contradictions are the most compelling realms of medicine; they take me back to the awe of my lungs lifting me through water. For me, as for centuries of doctors before me, my journey through these crossroads began when I first took a blade in my hand and cut a line across a dead woman's skin.

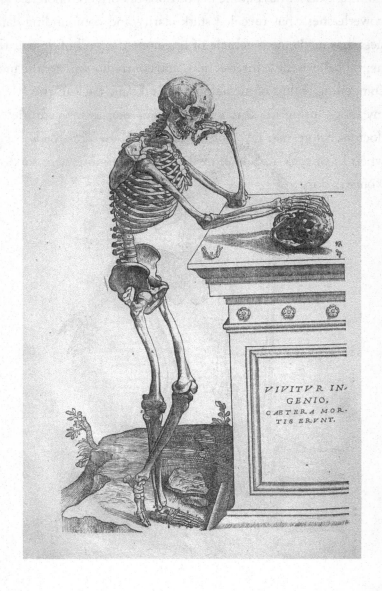

VIVITVR IN-
GENIO,
CAETERA MOR-
TIS ERVNT.

Bone Box

There we found a rich supply of bones, which we examined
indefatigably . . . until, . . . blindfolded, we could identify by touch
alone any bone which [our fellow students] pulled from the piles . . .
and handed to us. We were forced to these lengths because, though eager
to learn, we had no teachers to assist us in this aspect of medicine.

VESALIUS, ON STUDYING AT THE
CEMETERY OF THE INNOCENTS IN PARIS,
ON THE FABRIC OF THE HUMAN BODY

The syllabus says, "*Week One—5 P.M. Pick up bone boxes.*" The anatomy lab is empty, and yet it's just eighteen hours or so before our class. The cadavers have been prepared for months, if not longer, and as we will learn early in the term, their formalin embalming would permit them to sit at room temperature for as many as twenty years without the slightest trace of rot. They could be lying here awaiting our morning dissection, but the faculty is administering this intimacy with death in small doses. The absence of dissecting tables and bodies allows us to resume our social banter, to continue to introduce ourselves to one another before we undertake our strange new trade in the morning. Since the semester has not yet officially begun, it is the first day that we have all formally gathered, and we've come straight from an afternoon barbecue on a university playing field. Though many of my classmates know one another from their undergraduate years, I am meeting almost all sixty-some of them for the first time. Our conversations are friendly, a jumble of small talk about how we've all just moved and are still unpacking our belongings, polite questions about what we were each doing before medical school and what fields of medicine we think we might pursue. *My partner, Deborah, and I were both teaching high-school kids in California,* I say, over and over. I repeat the same joke: *So I'm obviously interested in psychiatry.*

Our talk is frivolous, but I'm aware that this moment is the real beginning of what will be four grueling years of work and study. Deborah and I moved across the country for this day. (*Not only have I never been to Rhode Island,* I said in a moment of uncertainty after the decision had been made, *I don't even know anyone who's ever been to Rhode Island.*) And yet here I am, with my new classmates, in an otherwise empty human anatomy lab.

We form a line, pose for Polaroid photographs that will help the professors associate our names with our faces. After the pictures we each pick up a wooden, handled box a little larger than a briefcase. The box has my name imprinted on a label by the handle. It is the kind of box that, in an academic community like ours, would be assumed to hold a telescope, or microscope, or collection of fragile documents. It will hardly be noticed, I discover, as I walk down the main drag of campus, past the falafel joint, the copy shop, and the Starbucks, carrying two-thirds of a human skeleton.

The box bumps against my leg as I take it to my car. During the entire walk, I am thinking, *This used to be a person. I am carrying parts of a person in this box, and no one knows it.* On the street, girls compliment one another's shoes, and a man in his twenties sings Dylan on the curb for quarters. When I reach the trunk, I hesitate for a minute and wonder if I should put the box on the seat beside me instead, and then I decide I am being ridiculous. I do not look at the box again until I have been home for over an hour, have unpacked my weighty new books and arranged them on my shelves. I do not look until there seems nothing left to do but unfasten the clasps and lift the lid.

Inside is a whole skull, at once eerie and beautiful. On close inspection the individual bones of the skull are visible, and their lines are fluid and lovely—the winging curl of the zygomatic bone that can be traced from the cheekbone to the ear, the bony hinge of jaw, the whorled external acoustic meatus, through which sounds travel to our brains. The lower jaw, or mandible, is held in place by small wire springs and screws and can therefore be carefully opened and closed. The top of the skull has been sawed off and reattached with removable fasteners to allow study of the internal cavities and structures of the head.

Incomprehensibly, there are twenty-eight bones in the skull alone. I will learn that there are fourteen bones that form the face, six auditory ossicles—the bones of the ear—and eight bones of the cranium, the "vault of the brain." The bones join together at irregular lines called sutures, and the look of them is half that of sewn fabric, as the name implies, and half plate tectonics; the ragged and uneven lines imply an ancient and irreversible joining. When I unhinge the top and look inside, the landscape is not the glorified basin I had imagined—it is lunar, with deep pits and sharp protuberances. Symmetrical sets of holes mark tunnels through which nerves and arteries once traveled, carrying signals and sustenance. At the base is the foramen magnum, literally "big hole," where the brain stem and spinal cord emerge from the brain to run down the body.

The skull has a full set of teeth, and one of the incisors had been crowned, as I can see from the whittled shape. I think, *Someone kissed this mouth; someone touched this chin in love.*

I lift from the box one of each of the bones in the arm—humerus, radius, ulna—and one of each of the bones in the leg—femur, tibia, fibula. I hold in my palm the bones of one hand, then the bones of one foot. My skeleton is male, I am sure. The bones and feet and hands are long. If you fit the bones together, you learn the size this man was. At the elbow the radius and ulna meet the humerus in perfect, well-worn curves and grooves. I hold the three bones in place in a straight line, then bend them at the joint. When I do this, I am unnerved and put the bones down on the floor. The movement is utterly human. Unquestionably so. I look at them, now separated on my carpet, and think, *Bones, just bones.* I pick them up, join them again, raise the lower arm from where the wrist would be toward the absent shoulder, and they are not bones, they are an arm. The proportions are exact, of course, the movement something I have never noticed or defined before, but something I innately know—the movement that a human arm makes. The flex, the bend and swing, the slight inward turn toward the body, even when straightened and at rest.

A whole spinal column lies in the box beside the skull, each vertebra strung with fishing line in its proper order to form a chain of bones. When held aloft they form a skeletal silhouette of this person's back. The column strikes me as prehistoric-looking, reminiscent of the bones of large fish that hang in natural-history museums. At first the vertebrae look identical, a stack of round disks, their bony prominences fitting neatly into one another. But in fact the shape of each vertebra differs slightly from those above and below it, and to run the eye down the full column is a bit like watching a time-lapse film of a budding flower or a developing fetus, a gradual metamorphosis from one distinct shape into another.

A single hip bone sits to one side. The bone is odd and asymmetrical alone, only half of the pelvis, unable to form the perfect bowl that holds the entrails. Yet it is elegant and curving and alien. Its inner lines rise and flare and become its outer edges, like the bloom of a calla lily. A fistful of ribs nestle, curved into one another, and they look thin and fragile and almost translucent. The final two bones in the box are responsible for some of the most striking bodily shapes: the clavicle and the scapula. The clavicle is a nondescript bone the size of a thick pen, and it is hard to imagine that it traces the insistent line reaching from a woman's throat out to her shoulder. The scapula, or shoulder blade, is winglike and twisting. Its graceful shape barely alludes to the way it firmly roots the shoulder muscles and the upper arm.

Beneath all of these in the box is the patella, or kneecap, a large, misshapen lentil that has the singular distinction of being a sesamoid bone—a bone formed in response to shear forces within the tendon that surrounds it.

The bones of the hand and foot are also held together by fishing line to show how they "articulate" or join. The number of small bones in the hands and feet exceeds all the rest in the body combined, and the names, I will learn, are lovely and evocative: distal phalanx, capitate, scaphoid, triquetral, hook of the hamate. Some etymologies seem comprehensible: The lunate is moon-shaped if you squint a little; *pisiform* means "pea-shaped" in Latin. But as the semester wears on and we peer through flesh at these little bones, look at them on black-and-gray transparent sheets of imaging, hold these bone hands in our own hands and memorize the shapes and names and the muscles that move them, we will hypothesize and confabulate, in seriousness and utter lack thereof, about those that are less self-evident. Perhaps the hook of the hamate, a protuberance of one of the hand bones, dates back to the stigmatized descendants of Ham in the Bible; perhaps the styloid process, a bump in the wrist, denotes the exact spot where stylish women wear bracelets. Learning the nomenclature will become a vast game of memory. And as I sit on the couch with these bones whose names I do not know, all I think is that, of everything beside me, the teeth are what make the body seem the most real.

Stacked beside me on my sage green couch: this spinal column that wraps into a coil without muscle to hold it upright, hands and feet tied together with floss, this skull hinged and empty. A man's teeth.

I stand, and with my right hand I hold the knobby end of the skeleton's femur at my left hip. With my left hand, I join the tibia to the femur where the knee cartilage would have been. I am comparing his body to my own tall form. His legs are shorter than mine. His ribs too narrowly curved to wrap around me. I hold the threaded bones of his left foot against my right sole. Our feet are the same size.

Here is what I will learn: The most alarming moments of anatomy are not the bizarre, the unknown. They are the familiar.

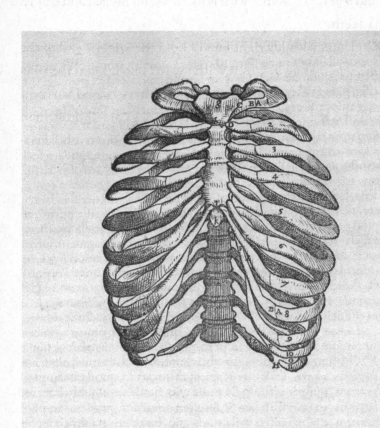

First Cut

*The body dead is, in a way, our world's greatest secret. We see always
flesh in motion, animated, disguised beneath its clothing and uniforms, its
signals and armatures, its armor of codes and purposes. When do we look
at the plain nude fact of the lifeless figure? Pure purposelessness—and
thus, in the absence of the spirit, strangely and completely present. Never
having a chance to see it, to assimilate our horror of it and go on to actually
look, how would we know that the lifeless body is beautiful?*

MARK DOTY, *HEAVEN'S COAST*

In the morning we have our first lecture before going into the
lab. I'm full of nervous energy. I've been anxious about dissecting a body since I decided to become a doctor, and sorting
through the bone box has only made that anxiety more real. I wake
up before my alarm sounds—rarest of rare occurrences in my life—
and am unable to quiet my mind. I turn on the coffeemaker, but
when I return to fill my mug, I realize I forgot to put the beans that

I ground into the filter. Clear hot water fills the carafe. As I walk across campus to the medical school, my mind keeps turning to the dead bodies I have seen at funerals, stiffly clothed and made up, and to my surprise at watching those mourners who reach out to touch the bodies, kiss their cheeks, and hold their hands. If those touches of kindness have unsettled me, how will I respond to the actions I will be asked to perform?

As my classmates and I greet each other and fill the lecture hall, I try to project comfort and nonchalance—as if, with my new pens and empty notebooks, this is any other first day of school. In fact, I'm full of jitters and wonder how I'll be able to focus on an entire morning of lecture.

Our professor, Dr. Ted Goslow, brims with enthusiasm at the start of this new semester, and he speaks for an hour and a half on the intricacies of the thorax—the section of the body between the neck and the abdomen that is enclosed by the ribs and contains the heart and lungs. "When you get just below the skin," he explains, "you're going to see two sets of important muscles, the pectoralis major and minor, or 'pecs' as you bodybuilders may know them." The class chuckles obediently, and Goslow continues, explaining what we'll encounter as we uncover layer after layer of the chest. "Between the ribs, three layers of intercostal muscles work in conjunction with the diaphragm to expand the rib cage and fill the lungs with air." Once we remove our cadaver's rib cage, he tells us, we'll see the lungs, and beneath the lungs we'll find the heart and its great vessels. We must cut these with care, he warns, in order to remove and study the heart. "By the end of today, you'll hold a human heart in your hands. It's amazing!" Goslow says. Until this point I had been diligently taking notes in the first pages of my fresh notebook, but as he conjures this image, I stop and sit quietly, a little slack-jawed.

Suddenly we are talking about reaching the cadaver's heart and lungs, and I have barely begun to get used to the idea of the initial cut through a dead body's skin. Friends of mine who'd finished medical school had alluded to the pace of the course when I asked about acclimation. "You don't have much time to process," one friend said. "There's just too much to get done." This class would be a baptism by fire. We would be given driving directions and a car with no brakes. Twice a week we would spend seven hours a day in the anatomy lab, where we could make right turns or wrong turns, but we would most certainly be moving. Dr. Goslow's tone is encouraging, yet also utterly straightforward: "This will be fascinating, frustrating, and technically and emotionally difficult, but it will also get done. We don't have much time, so get started."

Led by Dr. Goslow, my classmates and I travel en masse to the anatomy lab from the lecture hall and enter a narrow hallway filled with rolling racks on which hang scores of white coats. An additional group of instructors is waiting for us in the hallway— a mix of graduate students and junior faculty members who will supervise our dissections and provide us with much-needed guidance. One of the instructors tells us the three-number sequence to the combination lock on the wide metal set of double doors to the lab hallway, numbers we will all punch in at various hours as we come in late or early over the course of the semester.

Immediately after the announcement of the combination, I'm too nervous to recall a single one of the numbers. Already the sharp smell of the formalin-and-alcohol embalming fluid is washing over us. We try to ignore the smell, as well as the doorway that opens into the lab at the end of the hall. Instead we focus intently on the sizes of the white coats. None of us seems to be able to find the right size. We laugh nervous laughs, until we finally manage to slide our arms into jackets whose cuffs reach our wrists and there is no longer any reason to linger in the hall.

The view into the lab is jarring—eighteen white body bags atop stainless-steel rolling tables. There is no way for me to mistake the forms for anything other than human, but the fact that the bodies are enclosed and airless also makes them unmistakably dead. As we enter, we are supposed to grab our name tags off the table to our right and clip them to our jacket pockets. I forget and head mechanically toward the table to which three of my classmates and I have been assigned during lecture. I am hoping for a woman. Perhaps I want to learn about my own interior. Perhaps I think I cannot take anything that makes this experience more foreign than it already is.

I try to assess the form on the table without touching it, only looking at the way the thin, white, zippered plastic bag encases it, and I decide it must be a female form. Many of the bodies are unquestionably male, due to postmortem erections that make an odd tent shape of the bag. I have heard jokes about anatomy groups naming their cadavers "Woody." At the moment it is hard for me to picture—I feel far from being able to joke about any of this.

You'll hold a human heart in your hands, Goslow had said. But by the time I enter the lab, I have forgotten the promise of discovery and am focused only on his more practical advice: "Some reflex pathways we can control, like our eye-blink reflex, and some we can't, like our knee-jerk reflex. This explains that our brains sometimes overpower our wills. So if, once we go down to the lab, you see someone talking but you can't hear them, sit down, because it means you're going to faint."

I am the first of my group of four to arrive at our table but am joined quickly by Tripler, a bright and wonderfully quirky ex-ballerina; Tamara, a shy and often-absent twenty-one-year-old whom I will know little better on the last day of lab than I do on the first; and Raj, a recent biology major who cannot wait to begin dissecting. Like Tamara, Raj is twenty-one and has come to medical school straight from college. Tripler and I are both twenty-eight, having followed circuitous paths to med school. A no-nonsense intellect, despite her swinging blond ponytail and chirpy voice, Trip studied at Harvard and Oxford and has a master's degree in the history of medicine, which allows her to pipe up with often obscure but always fascinating historical tidbits over the course of the term.

"Listen to this, you guys," and some ridiculous but interesting story will follow. "In the 1700s this English country doctor thought that the dairymaids in his town might be immune to smallpox because they were lovely and smooth-skinned despite their constant exposure to cowpox. So, to test this, he inoculated some poor kid with the cowpox pustule of a dairymaid. Six weeks later he injected the kid with smallpox, and he didn't get it! Hence the first vaccinations!"

In the first nervous moments of lab, however, even Trip's merriment is muted. All four of us surround our table while other groups are still trying on coats and looking for their table numbers. Tripler and Tamara and I are quiet, our gazes bouncing from the body bag in front of us to one another, to the hushed and filling room. Raj is pulling small tools out of a drawer, chattering about dissections he's done in his biology classes: fish and lambs, piglets and cats. He is the only one who is excited to begin.

Once all the students have gathered around the eighteen tables, the instructors disperse among us to demonstrate dissection techniques. Dr. Dale Ritter comes to our table. He's an affable and frighteningly knowledgeable young member of the faculty, with an easy laugh and an unmistakable southern twang. When he unzips the bag covering our cadaver, we discover that she is indeed female. Her torso is covered with a damp white cloth. Her hands, feet, and head are wrapped in a translucent, cheeseclothlike material and then enclosed by tightly tied plastic bags. This elaborate wrapping, Dale explains, is to protect those parts of the body from desiccation until we begin our study of them. He adds that their coverage also helps depersonalize the body. The hands, feet, and head are parts of the body that are instilled with character. They can most quickly conjure up an individual life. But I cannot take my eyes off the woman's arms. They are covered in age spots and thin and long. They have skin that drapes from the bone. They are surely the arms of an old woman who has spent time in the garden or at the lake. They are the arms of my grandmother, which I massaged for a week before coming to medical school as she lay in bed following a stroke.

I see that she is not my grandmother when Dale pulls back the sheet to reveal gray hair matted between the woman's legs. In

March my grandmother had laughed and bragged to me that "one of the best things about getting old was not having any body hair." She didn't have "a speck of pubic hair" left. Was this woman younger than my eighty-year-old grandmother? I wonder. Was her death expected? Over the course of the next few months, would I learn things about her history from looking at parts of her that no member of her family had ever seen? Sutures from an accident, an empty space where her uterus should be, atherosclerotic arteries, a missing gallbladder or appendix? I would certainly learn some things about her that they would never know: the exact shape of her stomach, the marks left by pollution on her lungs, irregularities in the paths of her veins, the precise heft of her brain, the look of the inside of her eye.

Dale tells us we must spray our bodies with a wetting solution to keep them from drying out and making dissections stiff and difficult, if not impossible. He is all business. I feel as though I'm in two places at once. I listen intently to Dale, relying on his practical words to ground me in this most preternatural of moments; however, I'm also staring at the cadaver, wondering how I—how any of us—will be able to make the first cut into this woman's body.

As it turns out, by either coincidence or design, it's a leap that we do not have to make. Dale props our cadaver's right shoulder up on a rectangular block of wood and pulls a bright ceiling lamp over her to light her upper arm. "I'm just going to do a quick demonstration of dissection technique," he explains as he cuts a confident six-inch line and then perpendicular three-inch lines at either end of it, making a wide H. Then he pulls the skin up and shows us the fascia beneath, a yellowy, loose, fatty, connective tissue. We will discover that it is at times thick and greasy and globular, at others webby and thin. Below it we see muscle, gray and well defined, with clear linear fibers that make it look more ordered and purposeful than the fascia overlying it.

As Raj has discovered, several instruments are kept in drawers at the head and foot of each metal table, and Dale shows us how to use the scalpels to cut quickly and easily through skin. So sharp is the scalpel that you practically trace rather than cut with it. It is immediately evident how fast and irreversible the scalpel is, and Dale turns to slower, more cautious tools, which will therefore be those we choose most often. Scissors are rarely used to cut but are much more often inserted, closed, into spaces beneath skin or between arteries, or veins, or organs, or nerves, and then opened, to spread away fascia and connective tissue, separating out the structures intended for study. Two types of forceps are available, one with thin, blunt ends and one with a "rat-tooth"—a metal V on one end that fits into a corresponding notch on the other—to grasp and pick away extraneous matter.

We wear latex gloves, and Dale notes that, more often than not, our hands and fingers will be our most valuable and effective tools. He slides his finger into the incision he has made and then beneath the skin. Then, with remarkable ease, he makes an authoritative sweeping motion, which divides the layer of skin and fat he now holds between his finger and thumb from the gray muscle below it. He peels the skin away, revealing a section of overlapping muscles of the upper arm. The emergence of the muscles is an introduction to the promise of discovery, of clarity beneath disorder.

Before we begin to try these techniques ourselves, each group is called into another room for a prosection, a demonstration of what we are instructed to uncover in our impending dissection. This will be a far more elegant version than our efforts will yield. The demonstration cadaver is the body of a big man with tattoos on both biceps, which have blurred into large blobs of inky blue. The skin of his chest is gone. We watch Dale peel back the man's pectoralis major and minor, then lift off a large, precut square of rib cage revealing gray-blue lungs and a reddish diaphragm. He explains that we will use a saw to cut through the ribs. The saw, like the light, pulls down from the ceiling. I can see through the window to the lab room that groups who have left the prosection are already beginning to saw. Thin curls of cartilage and muscle are wheeling off the blade, and the dust of bone is breaking the light in the room into rays. The dust in the air smells like the dentist's office. I am eight years old and have fallen mouth-first into the concrete, with my friend Myla Wilson on my back. The dentist is grinding my teeth down.

I am afraid that I won't be able to do this.

The *Essential Anatomy Dissector*, our textbook version of a road map for the term, reads simply, "Make the incisions shown in Figure 1.1." The figure is a drawing of a naked woman, apparently in her thirties or forties, who looks straight off the page with open eyes and appears to be alive. She is, however, conveniently in what we have learned is called "anatomical position," which means that for this dissection she lies faceup, with her arms at her sides and her palms facing the ceiling. She has a full head of hair, which our cadavers, we can tell through their shrouds, do not. Their heads have all been shorn, giving them a look that is part androgynous and part awful, like prisoners of war. When I mention this to a friend who is already a doctor, she tells me that the cadavers in her lab still had their hair. Imagining how the hair would look after weeks of wetting solution and bone dust and how, early on, it might make them seem more human, I prefer our cadaver as she is.

My sense of the humanity of our cadavers is evasive and shifting. One of the strangest things about dissecting a human body is the difference between a human body and a human being—in some ways readily identifiable and in others barely perceptible. Everything tangible that is human is present in our cadavers. Their dead body parts are structurally identical to our living ones. Our cadavers are undeniably human. Each bears distinguishing traits that evoke the life of an individual. Once her hands are uncovered, we learn that the woman at another group's table has lavender polish on her fingernails, and, as a result of that lone variation, we find ourselves wondering about her: Did she have a weekly manicure appointment that she never missed over the last twenty years, for vanity and to share confidences with her manicurist? Did she live in a nursing home, which had a Beauty Day that she initially resisted, eventually acquiescing, only to love the experience of her first manicure but hate the old-lady color? Did her granddaughter paint her nails on her annual visits, despite the old woman's inability to recognize the girl after the ravages of Alzheimer's? Was the polish a preparation for a dinner date, an anniversary, a wedding, a funeral?

We can make cuts through our cadavers and peel their skin away. We can trace the paths of their circulatory systems and marvel at the fragility of vein and strength of nerve. We can curse the difficulty in finding a tiny artery in the thumb or neck and even laugh at our ineptitudes and mishaps. But the humanity of the body emerges in unexpected moments, and the balance of our voyage of discovery with the voyage of a finished life is sometimes difficult to steady.

Dissection, we will learn, will require us to turn off, in a sense, our connection with this humanity. I cannot say whether for me that is a conscious or subconscious decision, but I do remind myself more than once that my grandmother is alive in Indianapolis. I take a deep breath, and I focus—on the *Dissector,* on the scalpel and how much pressure is required to go deeply enough to proceed but not so deeply as to do damage. And I do this—this breathing and focusing and gauging and cutting—for hours. The work is slow and deliberate, and this first morning of lab will bleed into the afternoon and then the evening, with all groups progressing forward, little by little.

The skin of the chest pulls back easily after we have made the incisions, and the body opens like a book. Thumbs inserted at the midline of the chest above the sternum, or breastbone, pull back both sides, like the covers of a text, revealing the ribs and the muscles that connect them. In the female cadavers, the breasts, firm and set in an unchanging shape by the embalming process, remain attached to the skin. They are removed from the body when the folds of skin over the chest are peeled away and replaced when the skin is pulled back over our day's dissection. It is as if the breasts are part of a strange vest, as if this marker of gender can, in death, be slipped off or on.

When it is time to cut through the rib cage, none of my group members wants to do it. Neither do I, really, but I do, not to be macho, and not to force myself through something unpleasant or difficult, but because I need to know what I am capable of. I do not know whether I will be able to do it until I do. I am almost surprised when the body does not flinch or cry out when cut. And so, just as the humanity of our cadavers asserts itself in nail polish and tattoos, the inverse of humanity emerges in the body's utter lack of response to profound wounds.

Instinctively we find ourselves using our own bodies as reference points, as road maps. Perhaps due to her years of ballet training, Tripler stretches and hops absentmindedly in lab from time to time. When we begin studying the motions of muscles, she is brilliant. "What is the action of serratus anterior?" we ask her, already trying to quiz ourselves on the first structures we encounter. She is wrist-deep in dissection, separating out the pectoralis muscles, and instantly lifts her straight arm out to her side, then raises her outstretched hand above her head, saying, "It rotates the scapula when you raise your arm." We pause, holding our tools still above the cadaver in order to watch Trip's demonstration. "And it draws the scapula forward when you push your arm forward," she explains, miming a forward punch with one gloved and messy fist.

We begin this kinesthetic learning without knowing it: We are nervously trying to determine where to saw across the top of the rib cage, as the *Dissector* instructs us. Tripler reads to us from the textbook, "We go through the manubrium, so we have to find the suprasternal notch," and we all put our fingers to our own throats and find the thumb-size bony notch at the base of the neck, "and that's the top of it, and then we keep going down a couple of inches to the angle of Louis—you should feel a transverse ridge on the anterior aspect of the sternum, and that's where the manubrium ends and the sternal body begins." I've got the notch but can't find my angle of Louis, and so she has me feel hers, then try again. Ultimately she finds mine for me, putting my hand in exactly the right place. Dissection, we discover, is in part a process of beginning to name parts of our own bodies whose names we have never known. We find a structure beneath our own skin, a place we have toweled dry perhaps ten thousand times and never noticed, and then we uncover it in our cadaver, feel the shape of it, learn its purpose.

To cut away the front of the rib cage and remove it from the body, the *Dissector* tells us we must cut through, "ribs 2 through 6 just anterior to the serratus anterior muscle." I guide the saw down the body's sides. If you put your hands on your hips and then slide them up your sides to just beneath your armpits, your fingers are exactly where the projections of your serratus anterior muscles are. I am sawing alongside where both breasts were just hours ago, down to the sixth intercostal space—the space between the sixth and seventh ribs on both sides. I then saw across the body so that the two parallel cuts meet. To free the top of the rib cage, I have no choice but to go through the manubrium, which I can now locate easily.

I must cut with the bone saw through each rib and through the muscles and cartilage that support them. The saw is imprecise and requires a good deal of force—you have to lean the weight of your own body into it. Sawing through the rib bones is disconcerting; enough pressure must be applied to coax the dull blade through the bone, but too much sends it sailing past the far edge of the rib and into God knows what—whatever is in the space into which we're entering blindly. The muscle and cartilage are much easier to saw, but, as a result, doing so lacks the distraction that effort affords. The tissue spins off the blade in small bits, which look like tiny roots or fingernail clippings. The cutting doesn't feel awful or triumphant when I finish; it only feels done, and suddenly quiet, and, after all, there is now the first dark and open space to look into.

With the sawing finished, my lab partners can begin to separate the rib-cage section from the body, cutting through any remaining connections with their scalpels. In the center of the section at the top and bottom is a set of blood vessels that run along the inside of the breastplate. They must be cut before the rib cage is removed. When my partners do so, we can lift the rib cage free. Its outline dips down in the center to the midline sternum and then flares out at both top and bottom to either side with the rows of ribs. Disembodied, the rib cage looks like a butterfly.

Most of us, I think, harbor an ingrained, innate aversion to doing willful harm to the body. Of course, such harm is regularly done: adolescents shoot each other over vapid allegiances, men kick their pregnant wives, children smash beetles to bits under their heels. There is an internal restraint, however, that must be overcome, by rage or fear or sheer will, before we will do harm to a body. And so, watching the flesh coil away, the bone rise above the chest as dust, I feel that feeling you get when you think of the moment your teeth broke as a child or you hear about a fracture in which the splintered bone pierced the skin—an inescapable feeling of *wrong*. Tamara cuts through skin, and I feel the instinct to place my hand on the cadaver's arm: This will only hurt a minute. This will be over soon.

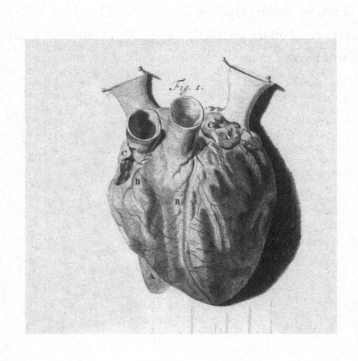

Breath and Blood

Anatomical textbooks give the misleading impression that everything in the chest is immediately distinguishable. The unsuspecting student plunges into the laboratory carcase expecting to find these neat arrangements repeated in nature, and the blurred confusion which he actually meets often produces a sense of despair.

JONATHAN MILLER, *THE BODY IN QUESTION*

The opened thorax, or chest cavity, reveals two large, gray-brown lungs. Though they are similar in size, they are asymmetrical by design—the right lung has three sections, or lobes, where the left lung has only two. The top of the heart is positioned at the body's vertical midline, directly between the lungs, beneath the sternum. The base, though, veers off to the left, appropriating some of the space into which the left lung could have otherwise expanded. The hearts of the cadavers in the room vary greatly in size. Ours is on the smallest end of the spectrum and is only a little larger than my fist. The heart of the cadaver at the table behind ours is, like the dead man himself, much, much bigger. With the fat that encases it, his heart is nearly the size of a Nerf football.

In order to examine the lungs and heart, we must remove them from the thorax. As with sawing the rib cage, some force is required to do so. We reach beneath the lungs to find the root structures—the forked bronchi, which divide the trachea's path to serve both the right and left lungs, and the pulmonary arteries and veins, which connect both lungs to the heart. The *Dissector*, in its typically obtuse tone, instructs us to "push the lung laterally to stretch the root of the lung and carefully transect the root structures." A great deal of effort must be exerted to stretch the root of the lung, and Tripler and Raj wince as they slide their gloved hands into the pooled embalming fluid at the base of the thoracic cavity. They flex their arms and pull up on the lungs from the connections at their base. I have the scalpel, and I lean over the table at different angles to try to see where to cut. With the lungs still encased by the lateral rib cage, and my partners' four forearms already filling the space, I must make a nearly blind cut. The trick is to sever the bronchi and the pulmonary vessels while preserving as much of them, and of the lung tissue, as I can. And I must exert care not to slice into the hands of my lab partners. Tamara buries her gaze in the *Dissector* and reads aloud the path each cut is supposed to take. When the cut is done, the lines of separation are jagged and imprecise, and too little of the pulmonary vessels remain in the thoracic space. Raj and Tripler have soaked the cuffs of their lab coats in the formalin embalming fluid that the lungs had absorbed. We are frustrated and decide to remove the heart after a break for lunch.

When we return to the lab in the afternoon, we have a bit of renewed energy for the remainder of the thorax dissection. Now that we have removed the lungs, the heart has more space around it, and once we have cut through the vessels to prompt its release, it comes out easily. But when it's time to dissect the heart itself, the process feels inexact and takes a great deal of time. The embalming fluid has combined with hardened blood to fill our cadaver's heart with something that looks strangely like over-cooked, processed chicken. All afternoon we are fishing this stuff out of veins, arteries, and heart spaces.

The heart muscle itself does not look so different from chicken—a little stiffer and grayer, perhaps—so, as we attempt to clear out the blood and embalming fluid, we cannot tell the difference between the flesh and the blood. We feel all the while as though we are actually tearing into this woman's heart. Occasionally we come across wide, thick tubes of hardened blood, and red-black crumbles of it cover everything. We try to run water through the vessels, and the pieces of blood clog the drain of the lab sink. The white cloth covering the body is now dirty and stained, as is the body bag, and the feeling of imprecision and carelessness is exasperating.

For reasons partially physiological and partially philosophical, only the brain supersedes the heart and lungs as the most essential human organ. The loss of brain function invariably leads to cardiopulmonary death, though several hours or days may pass in between. We know that brain death is synonymous with the loss of personhood, and therefore with the loss of life. Surely there will be a voice or two raised here objecting that as long as we can hook bodies up to machines that breathe air into them, and pump blood through them, and maintain the look and feel of a body, warm and alive, then life continues. I disagree. Scientifically I side with the neurologists who designate death as the irreversible cessation of all functions of the entire brain. Philosophically I wade into murky and controversial waters with personal beliefs that the boundaries of death are nebulous. To me the definition of "dead" may extend far enough toward life to include a person whose brain is a physically functional void: the end-stage Alzheimer's patient, for example, who never wakes but breathes and sends urine from injected liquid food into her diapers.

You could make the argument that a few other organs are indispensable. If the kidneys fail, we need dialysis to rid the body of its waste and toxins or else we poison ourselves, no matter how healthy the heart, lungs, and brain. The necessity of other failing organs has been bypassed by medical advances: If the stomach doesn't work, we can surgically implant a feeding tube that empties liquid nutrition directly into the intestine; if a cancerous rectum is removed, we can attach an ostomy bag to the patient's colon to collect the products of the bowels. The gallbladder is routinely removed in people prone to gallstones, with no severe repercussions.

The seventeenth-century English physiologist Richard Lower conducted a simple experiment that illustrated the essential role of the heart and lungs in living beings. Since ancient times it had been known that the blood entering the heart had a purplish-blue color to it, and that the blood exiting the heart was a deep, rich red. Lower demonstrated that this color change took place in the lungs. The question remained: Did the color transformation occur simply due to the blood's passage through the lungs, or was exposure to air integral to the change? In order to answer this question, Lower performed two experiments. In the first he pumped blood through a dog's lungs, which he hooked up to a kind of artificial respiration. As long as the "breaths" of the respirator continued, the blood that emerged from the lungs was red. However, if the respirator was stopped, the blood stayed blue.

The second experiment ensured that there was not some intrinsic aspect of the lungs that caused the blood to undergo its change. Into a vial Lower poured purplish blood, which promptly clotted. The top surface of the clot, which was exposed to air, changed to a vivid red, and the blood below the surface remained purple. When Lower removed the clot and turned it over, the colors inverted. The red faded from the portion of the clot that was now at the base of the vial, and the newly exposed surface of the clot brightened from a purple-blue to bright red.

With these two simple experiments, Lower established the collaboration of the heart and lungs in loading blood with oxygen and distributing it throughout the body. And it is this understanding of the precise purpose of the heart that guides our dissection and our study.

The language of the heart is lovely. Divided into distinct functional and structural quadrants, the heart is routinely described as "four-chambered." Tiny valves open and close—doorways into each chambered space—to allow or prevent the entrance of blood. The top chambers of the heart are the right and left atria. The bottom chambers of the heart are the right and left ventricles.

The right and left sides of the heart work in parallel to one another. The right side collects Lower's blue blood, drained of oxygen by the workings of the body, and pumps it back to the lungs for fresh contact with inhaled air. The left side is devoted to gathering the blood from the lungs once it is again heavily laden with oxygen and propelling it to the body's farthest recesses. On both sides the blood first pours into the delicate atria and then flows into the thick-walled, muscular ventricles, which contract mightily to send it to load or unload precious oxygen.

The atria fill simultaneously, then empty into the ventricles during a period of heart relaxation called diastole. The ventricles contract together in systole, sending deoxygenated blood to the lungs and oxygenated blood throughout the body. This emptying and filling, receiving and sending, happens more than once each second. The blood loops and returns, loads and unloads, empties and fills. Systole and diastole are marked by the two sounds of the heartbeat, and they each occur around seventy times a minute. When the body needs more oxygen—in exercise or in fear—the heart may sound 120 times in those sixty seconds. Maybe more. The beating of the heart is one of the first internal things about our bodies that we understand. And yet how can our hearts beat 3,153,600,000 times, give or take a few hundred thousand, over the course of a lifetime? How can they not predictably and catastrophically hiccup sometime in the first year of life with more than 42,048,000 beats in which to do so?

This is one of the experiences in the anatomy lab that I do not expect: When I look at the tissues and organs responsible for keeping me alive, I am not reassured. The wall of the atrium is the thickness of an old T-shirt, and yet a tear in it means instant death. The aorta is something I have never thought about before, but if mine were punctured, I would exsanguinate, a deceptively beautiful word meaning that my blood would spill throughout my body. This is a condition the textbooks describe in a blasé tone, as if they were talking about a fouled relationship: It is incompatible with life.

During this first long day of lab work, I have realized that Raj and I are polar opposites in our approach to dissection. I do not react at all the way I would have expected. Unlike in almost every other realm of my life, in lab I am tentative and cautious, perpetually afraid of cutting something important. But I also feel as though I should cut these things *well*. Though we each inevitably sail through an important artery with our scalpel, or remove a superficial nerve without even knowing it, I am aware of the fact that in dissection there is no going backward. Raj, on the other hand, moves quickly and with less concern. When we read in the *Dissector* that we are to cut into the right ventricle so that we can see the thickness of its wall, Raj makes the incision swiftly and with certitude and sort of shreds it. I am a little saddened by that, even though I'm not sure it was avoidable. Before long it becomes clear that neither of our methods is quite right. Raj charges ahead, sometimes inadvertently destroying a structure that we needed to preserve. Though our two styles will grow more similar as the semester progresses, I instinctively cringe while watching him cut, and he sits and sighs to watch me poke and check the book before I proceed. I am so careful that by the end of the day I am way behind schedule. We form a dynamic that will, unbeknownst to us, endure over the course of the semester.

Despite our stylistic differences, Raj and I have nonetheless aligned ourselves as the two lab members who are most comfortable performing the dissections. Tripler and Tamara direct us much of the time from the *Dissector*, point out landmarks, and ask questions that help all of us study. In most groups the roles sort themselves out in similar—and similarly unspoken—ways. Raj naturally gravitates toward cutting, as he has since the first moments of lab. Yet I am surprised to learn that even if the actions we are asked to do are distressing, I am more comfortable *doing* the dissections than observing them. When Trip takes over for me, I hover, restless and bored.

It is only the first afternoon of lab, and already, when Trip asks me questions, I am feeling behind. There are so many names of veins and arteries, so many tiny parts to the heart and body that we aren't even sure we have correctly found. Many more don't seem worth the effort to locate. We can easily see the superior and inferior venae cavae. But seemingly endless numbers of other landmarks are more like the oblique pericardial sinus, a space within the heart bounded by five vessels into which we stick our fingers and kind of shrug: *I guess that's it.* Whether or not it is, we move on.

When the first day of dissection comes to a close, Tripler and I head to our lockers. We chat absentmindedly as if today is no different from any other day for us, as if we haven't touched our first dead body—cut it, even, taken out its heart and held it in our hands. *Its heart*. The "it" is what bothers me.

"Trip," I say, "I just feel weird about . . ."

"The face," she says. "Me, too."

We walk back to the lab door, fumble in our notes until we find the combination. We are the only ones in the room, and it is strange all over again. When we stand above our cadaver as we have done for the last six hours, the room is silent. Twenty bodies in the room, only we two breathing.

"It doesn't feel right to cut her up without knowing what she looked like before," I say. The reiteration is to buy time. We won't see what she looked like before, of course. She is shorn; she is the gray-green of formalin embalming. We can tell even through the cheesecloth, through the plastic bag, that her mouth is open. Still, she will look less herself by the time we are meant to take off the bag for the face dissection at the term's end.

Trip nods. She is resolute. She unwraps a new blade for the scalpel and slides it onto the handle. She cuts the string that ties the bag over the head at the neck. I pull off the bag, unwrap the cloth.

We hold our breath.

The truth is, she is beautiful.

Fine features; small, angular nose; thin lips. A face capable of intensity. At her throat a dark braid of stitches closing the spot where her blood was exchanged for embalming fluid.

Her eyes are open. Her mouth is open. Tongue. Teeth. She looks as if she would speak if she were not this wrong hue.

I did not expect her to be beautiful.

I bring my partner, Deborah, to the lab that night. "It is so clear that these are not people any longer," she says. She is right. And she is wrong. The time eddies and swirls. They were people, are people, were loved, are loved, were bodies, are bodies, have died. And what else?

The syllabus informs us that we will spend two entire weeks dissecting the thorax, due to the sheer number of structures to consider. During the second week of these dissections, our anatomy instructors have gone to a butcher shop to get a heart, trachea, and lungs from a cow that was butchered this morning. The cow's trachea is wide, bright white, and plasticky, and it looks like the hose of a bilge pump. The heart is huge and a great relief, as it can easily illustrate the tiny structures that are so difficult to locate in our human specimens. We see the oversize tricuspid valve and the chordae tendineae, the fibrous cords that connect the cusps of the valve to the papillary muscles. When the heart contracts, the papillary muscles also contract to pull on the cords and prevent the cusps from inverting into the atrium. In our human hearts, the chordae tendineae are a thickness somewhere between thread and embroidery floss, but in the cow heart they are far less fragile. The cow's lungs are the most remarkable. At first they look like a small, pink, split manta ray. We touch them and they squish, like a sponge filled with water. Then Dale hooks them up to an air hose and they grow stiff instead of spongy and inflate until they are half of a huge beach ball. They swell to five or six times their original size, and their capacity for expansion is permanently fixed in our minds. Trip, Tamara, Raj, and I return to our cramped and difficult heart, to our cadaver's gray-brown lungs, spotted with blue specks of pollution, amazed at them.

Our replenished energy does not make up for our growing sense of falling behind, and when we finish the second week of dissection, we know we will have to spend additional time in the lab before we become responsible for still more material in the week that follows. To that end I return to the lab on Saturday morning to study with Lex, a friend from class. Lex is a sweet soul with a British accent and a persistently half-untucked shirt who alternates between thinking he should be a doctor and thinking he should fix up old cars. He is an insomniac who studied literature at Bennington, and so from time to time he'll stay up all night with books of poetry and will come in waving pages of Mary Oliver poems for me to read.

During our Saturday-morning study session, I am teasing Lex about being both a vegetarian and a chocoholic when we stumble upon one of the most wondrous moments I experience in lab. We are looking at the semilunar valves in the pulmonary trunk and aorta, and they look just like their names, like little half-moons that work passively and without musculature, unlike the tricuspid valve with its papillary muscles. The idea behind them is that they surround the lumen of the vessel into which the blood is being pumped from the ventricles—on the right the pulmonary trunk and on the left the aorta. If the blood flows forward, they are pressed closed against the side of the vessel, so as not to impede the flow. However, should any blood attempt to flow backward, it would catch on the flaps of the three semilunar valves and instantly fill and open them, thereby closing off the vessel. When full, each of the cusps presses up against the sides of the others, their joinings splitting the circle of the vessel's lumen into perfect thirds. To visualize the valves' function, Lex and I decide to take a heart to the sink and pour water through the pulmonary trunk toward the ventricle. The semilunar valves work like a dream, catching the water as sails catch wind, closing fast and preventing any leakage. It is astonishing, almost impossible to believe.

Anatomical Precedence

This old man, a few hours before his death, told me that he was over one hundred years old. . . . And thus while sitting on a bed in the hospital . . . without any movement or other sign of distress, he passed from his life.

And I made an anatomy of him to see the cause of a death so sweet.

LEONARDO DA VINCI, REGARDING HIS DRAWING
A DEAD OR MORIBUND MAN IN BUST LENGTH

In beginning the process of cutting apart a human body, we join a long line of medical students—and pioneers in quest of the body's mysteries—who plunged into this forbidden terrain. The history of human dissection is a long and tortured one, and as I take my place in it, I cannot help but wonder what it must have been like for the first scientists and artists who dared to think of looking beneath a body's skin. How did they go about *their* first forays into the dead? I want to know what it was like for these audacious explorers, who were propelled by the same promise of wonder I feel yet who ventured into territory not only forbidden but largely unknown. I want to understand in a deeper way the intensity of the taboo of dissection and the visceral aversion we feel even today to the violation of the human form.

Representations of the human heart and trachea have, it turns out, been found on amulets dating as far back as 2900 B.C. Dissections are not formally documented, however, until two and a half millennia later. In fact, the seminal triumphs and tensions of anatomical discovery took root in the blaze of the Italian Renaissance.

The Italian university town of Padua is the official birthplace of human dissection, and the town's renowned anatomical theater was built in 1594. Similar theaters were built across Europe, with particularly famous examples in Leiden, Paris, London, and Bologna. Though the theaters were originally built for teaching purposes within university medical schools, the dissections performed in them ignited curiosity first among artists and academics in other fields, and eventually in the community at large. In many cities certain dissections were designated for doctors and anatomists only, but others were open to the public. Tickets were sold, anatomical demonstrations were given, and oftentimes the events were preceded or followed by extravagant meals. Seventeenth-century travel diaries and postcards reveal that attending a dissection was a society event and marked a European traveler as on the progressive edge of culture. In eighteenth-century Bologna, as in many other Italian towns, dissections routinely took place during Carnival. The anatomical theater would be opulently decorated with candles and damask wall hangings. Masked attendees would fill the theater, and the dissecting table was illuminated by waxen torches placed at the head and feet of the cadaver.

The University of Padua, situated in the Palazzo del Bo, boasts a remarkable history, virtually unparalleled in academia. Italy's original university was founded in Bologna in 1181. In 1222, however, a group of students in Bologna grew frustrated with increasing restrictions on academic exploration—accepted by the university—imposed by the pope and the government. Padua's university was formed by these students as a place of independent learning, designed to remain unaffected by the ordinances of any pope, emperor, or king. This designation drew to Padua scholars from all across Europe who had felt constricted by regulations imposed upon their work. The list of names is mind-boggling: Copernicus; Galileo Galilei; the great British doctor William Harvey, who discovered the system of circulation of human blood; and many others.

Padua's university was the professional home of Andreas Vesalius of Brussels, the father of modern anatomy. In the twelve hundred years prior to Vesalius's seminal sixteenth-century masterpiece, *De Humani Corporis Fabrica* (literally, *On the Fabric of the Human Body*), anatomical study relied upon the writings of the second-century Greek physician Galen. Although Galen's own anatomical knowledge was derived from experimentation and observation, Latin translations of his texts were used in the Middle Ages as substitutes for dissection. When the rare dissection did occur, the anatomy professor would read Galen's writings, word for word, and would take no part in the dissection itself. In his book *Doctors*, the physician-writer Sherwin Nuland describes a fifteenth-century illustration of just such an occasion:

> The professor sits perched high on what is quite literally his chair, droning along in his recitation of the Latin Galenic text while an ignorant barber-surgeon dissects the cadaver below and a barely better-schooled demonstrator shows the body parts to the only mildly interested students. . . . Since the professor never descended from his magisterial throne to actually look at the structures being displayed, and neither the surgeon nor the demonstrator really knew what he was doing, the several days devoted to the exercise each year were little more than a walk-through to satisfy a curricular requirement whose advantages were more theoretical than real.

Vesalius, on the other hand, was a master dissector committed to the necessity of direct observation in anatomy. His *Fabrica*, as it is known, marked the first time in twelve centuries that the investigation of the body's structures had veered away from Galen's writings and returned to the body itself. In Padua, Vesalius was comparatively free to disobey the decree of Pope Boniface VIII that forbade desecration of the body and therefore outlawed dissection.

Vesalius's writings were accompanied by a glorious set of wood-cut engravings by Jan Stephan van Calcar, who studied under Titian. In the most famous series of the Vesalian illustrations, called "the muscle men," male figures stand in classical poses amid the Roman landscape. Their bodies are in various stages of dissection, but their positions and expressions are as lifelike as yours and mine. In some, flayed muscles hang from joints and bones, and yet the men gaze and ponder, stretch out their partially dismembered limbs. The woodcuts are strange and gorgeous, and in them life and death overlap to so great a degree that they collide.

Vesalius was audacious, and a man of insatiable curiosity. This ferocious commitment to elucidating the mysteries of the body is perfectly illustrated in one of my favorite passages in his *Fabrica*, as he describes his own acquisition of a skeleton. At the time a skeleton was a rare and precious possession, which Vesalius could not afford. In order to obtain one for his own study, Vesalius and his friend Gemma walk "in the hopes of seeing some bones." Their destination is "the place where, to the great advantage of students, all who have suffered the death penalty are displayed by the public highway," and there the two come across the skeleton of a condemned man, which had been hanging for more than a year. The criminal, Vesalius writes, "had provided the birds with such a tasty meal that the bones were completely bare and bound together solely by the ligaments." But, having found a skeleton, "such an unexpected and long-sought opportunity," Vesalius must now undertake the gruesome—and illegal—task of pulling it down and transporting it back to his home. With the help of Gemma, the anatomist writes that he "climbed the stake and pulled away a femur from the hip bone; and, when I pulled at the upper limbs, the arms and hands came away bringing with them the scapulae." Vesalius made several trips to and from the corpse, taking the legs and arms home in "several secret journeys," and, for the retrieval of the trunk and head, he writes:

I allowed myself to be shut outside the city at nightfall; so keen and eager was I to obtain these bones that I did not flinch from going at midnight amongst all those corpses and pulling down what I wanted. I had to climb the stake without any assistance, and it took a great deal of effort and hard work. Having pulled down the bones I took them away a certain distance and hid them in a secret place, and brought them home bit by bit the next day through another of the city gates.

Once in his home with the criminal's bones, Vesalius softened the ligaments in boiling water, cleaned the bones, and built the skeleton, replacing a missing hand, foot, and patella "with considerable difficulty from another source." Once finished, he was well pleased with his accomplishment, and brags a little: "I prepared this skeleton so quickly . . . that everyone was convinced I had brought it from Paris in order to avoid any suspicion of body-snatching."

Though he may once have legitimately feared the legal ramifications of his research, as his academic status and reputation grew, so did his boldness. Vesalius was invited in 1540 to conduct Carnival dissections in Bologna, the accounts of which marvel at the large supply of subjects made available to him. Six live dogs and the corpses of three executed prisoners were dissected in front of the crowds. The arrangement obviously allowed Vesalius some influence over the legal authorities, because after five days of lessons he confidently announced that "tomorrow we shall have a new subject; I believe they will hang another; indeed this cadaver is now dry and wrinkled."

During my first weeks of medical school, my friends and family phone to see how things are going. They all want to know about my first day in anatomy lab: *What was it like?* they ask. A mix of curiosity and horror in their voices. Some of them ask whether they can accompany me to the lab: Our friend Welly is an artist. She comes in with me late one night and tearfully sketches in her journal, moved by the matter-of-factness of the cadavers' mortality. Deborah's college roommate, Caroline, and her fiancé, Stephan, ask to visit the lab one Sunday. Stephan, who is a philosopher writing on issues of personhood, spends nearly an hour with me in the lab looking closely at the structures we've dissected; Caroline, an unflappable optimist and a ceaselessly inquisitive spirit, gets as far as the door, sees the body bags, and bolts out of the building. None of their reactions is unusual among the visitors I bring to see the body I'm dissecting—in fact, all of them are manifestations of the range of feelings that sweep over me from time to time through the course of the semester. But regardless of their reactions, all the visitors say to me at some point that the experience was not at all what they had expected.

Through visiting the lab, my friends and family members become more able to share in and understand this stage of my medical training. As more and more questions arise for me about what Vesalius and his fellow pioneers encountered as they ignored the established Galenic tradition and raced ahead in exploration, I long for this kind of firsthand view of anatomy's history. *What was it like?*

Therefore, when a summer break allows me to do so, I decide to travel to Vesalius's Italy. In search of a different view of gross anatomy from what I have found while working in the lab, I will visit the sites that set the historical precedent for cadaveric dissection and thereby revolutionized the basis of modern medical teaching.

I travel first to Padua and arrive at the Università degli Studi di Padova before it is open to visitors. As I wait, I sit on a bench adjacent to the entrance to read my guidebook and eat a handful of tiny strawberries that I've bought from the market for breakfast. At first blush the university does not seem like the serious foundation for anatomical study that it is. Just beyond the portal of Padua's college palazzo at 10:00 A.M., a twentysomething woman is standing on a public bench in the square. She wears a pair of large white briefs and a black bra, with a small white tank top sliding off her shoulders. She holds a bottle of champagne and reads aloud from a hand-drawn poster with a caricature of herself in the center and text all around it. As she reads, her friends smear her body with butter, wrap her arms and legs in plastic wrap, sprinkle flour in her hair, crack eggs over her head, shove dried pasta down her underwear, drape her with strings of sausages. She laughs. They laugh. At the periphery, middle-aged men and women stand in dressy clothes, giggle, take pictures. Every few minutes they all break into rounds of song. A copy of the poster is taped to the university wall.

Above the caricature, which on closer inspection is the woman, dressed in a bikini, leaning against a giant penis, is a title: *Monica, Dottore Filosofia.* Monica's reading goes on, and another woman walks through the portal in a gorgeous long beige skirt and fitted jacket, with stiletto heels. She wears a beribboned wreath around her neck, and her family and friends encircle her, beaming, and applaud. Street musicians strike up the song that Monica's friends have been singing, and in the portal the woman in the beige suit begins to dance, first with one family member, then another, one friend, then another, and then the crowd forms a bridged passageway with their arms, which she walks through. She is systematically kicked in the rear by everyone. Then she is led, beige skirt and all, to an adjacent bench where she strips down to her underwear, takes hold of a bottle of champagne, and is subjected to a similar fate as her neighbor.

The procession continues like this until lunchtime, until ten or more naughty posters line the university walls and as many young graduates, shaving-cream-clad with lipstick tattoos and bright wigs, leopardskin wraps, and athletic supporters, have led their families out of the square, presumably off to graduation lunches and a day of celebrations.

It is through these festivities that I walk to try to find a third-floor office, to ask about Padua's historic anatomical theater. As I speak just French and English, along with a few bits of medical Spanish (it is useful only in very specific situations, to ask, "Is there blood in your sputum?" or to say, "We'll need a sample of excrement"), I thumb through my Italian phrasebook as I climb the stairs. When I find the right door, I knock, and a voice calls out something in Italian that, to me, is equally likely to mean "Come in," "Wait a minute," or "Go away." I turn the handle and open the door, hoping an apologetic smile will compensate for a bad guess if need be.

"*Si?*" the woman behind the desk asks.

"To write. Book. Anatomy. Please. Permission. Photograph. Theater."

She looks at me quizzically, then takes my arm, and we cross the hall into a room of computers with two women inside. Italian is exchanged rapidly. For a long time. Then a young woman rises from behind a computer and walks toward me, hand outstretched.

"I'm Anna," she says. "How can we help you?" Anna, I will learn, is a stroke of great luck for me. Many times over I will make a request, like permission to photograph the anatomical theater. Anna translates this to Eleanora, the first woman I encountered. Then a five-minute conversation ensues in Italian that sounds every bit to me like an argument. Eleanora says numerous things fast, all of which sound like no. Anna responds. They do this for some time. Then they both turn to me and smile.

"It will be okay. Come with us."

We leave the office space and enter into the Palazzo del Bo. Anna and Eleanora and I climb the stairs of the ancient courtyard where Vesalius once studied, toward the door that will lead us into the "anatomist's kitchen," a room adjacent to the anatomical theater, where corpses were once prepared for dissection. The story goes that, despite severe criticism from the Vatican, the University of Padua was given two corpses per year by the liberal Venetian government. But they decayed quickly and were insufficient to support the students' needs. The students and faculty members therefore stole corpses from cemeteries. Legends of how this thievery was kept secret vary, but most agree that the table at the center of the theater flipped over, so that if inspectors or officials of any kind approached the theater, the table could be quickly turned so that the corpse was hidden, and a flayed animal would appear in its place.

One version of this story holds that the professors and students dug a tunnel under the dissecting table, which led beneath the walls of the city to the river. At night, groups would leave the city walls and paddle rafts to the graveyards, where they would dig up bodies and float them back down the river to the mouth of the tunnel. They would carry the bodies through the tunnel and, using a system of pulleys, lift them up to affix them to the underside of the dissecting table as needed. The lack of corpses was so severe that in the great oratorical hall adjacent to the theater a case holds seven skulls lined up in a row. The skulls are those of anatomy professors, who left their bodies to be dissected.

Unlike the decorated theaters of Carnival, the anatomical theater at Padua, with its rising, concentric circles from which the students observed dissections, is believed to have served a purpose more functional than spectacular. Today it can be visited on an official university tour. You enter the theater at its lowest level. Though it has been recently restored, no one is permitted to enter the levels once occupied by students. Positioned to look up into the rings of railings in the unique anatomical theater in Padua, one occupies the same space so many cadavers did. It is impossible not to imagine the faces of students peering around one another, over shoulders, into the center of this remarkable room.

There were no seats in the theater, so students stood with a rail in front of them and the rail belonging to the higher row behind them, close enough that both rails touched their bodies. For many reasons—reasons far more legitimate than ours on our first day of anatomy lab—students in the Palazzo del Bo's anatomical theater were likely to faint. The quarters were incredibly tight. The theater is tiny, but anywhere from two to three hundred students are said to have packed into it at a time. However, the small size was effective. It helped students to see as well as possible the corpse being dissected. It also allowed for easy communication and collaboration between students; those on one side of the theater could describe to students on the other what they were seeing.

In addition, there were no windows in the room until 1844, so not only were lectures held by torchlight but the smell of the (often already partially decayed) corpse must have been terribly strong. For many years dissections were held only in the winter, in an attempt to curtail the rapid rot in warmer temperatures. The hope was that the close quarters crammed students into the rows tightly enough so as to prevent them from falling forward into the center of the theater in the event that they should faint. Still, a diary entry from one anatomy professor reports that during a single lecture ten students fainted and fell from the rows. Such an occurrence was not uncommon, and students often injured themselves rather seriously in this way. As a precaution two measures were taken. First, it is said that orchestras were hired to play music for the students, to calm them as they entered the theater. And second, and more practically, that same professor calculated the railing height necessary to keep a body from falling forward and as a result raised all the railings by twenty centimeters.

The precipitous incline and sparse décor that surrounded dissections in the Palazzo del Bo differ from the feel of the anatomical theater in nearby Bologna, built nearly half a century later, in 1637. Rather than having the sharply pitched, circular rows of Bo, the room in Bologna is large and square, and less dramatic. Dark, wooden, backed benches are arranged in rows forming a cantilevered rectangle around the marble dissecting table, one of only two indicators that this is anything other than a picturesque old lecture hall. The other is the stately lecturer's chair, suspended along one of the room's walls. The chair is not unlike a throne beneath a wooden canopy. Two famous skinless statues of men, called *spellati*, effortlessly support the canopy, lifting it above their heads. All the while they demonstrate their own musculature, as if holding a great piece of wood aloft is the most natural thing in the world to do after having been skinned.

Standing in the core of Padua's restored theater, considering what dissections here must have been like, I realize I might easily have lacked the commitment to anatomy that the students in the sixteenth and seventeenth centuries had to have had. Not only did they engage in grave robbing, but they had to be willing to defy papal edict, either solemnly or with a bit of jest and pretense.

This apparent mockery of church beliefs manifested itself officially in the great hall of the Palazzo del Bo, where to this day a painting above the hall's entrance designates patron saints for the two original divisions of the university. On the left, Saint Catherine of Alexandria is depicted as the protector of the students of law. On the right, the protector of students of medicine and liberal arts is none other than Jesus Christ, the Redeemer.

Yet despite the brazenness of some university figures, who must have deemed ridiculous the Vatican's demonization of scientific discovery, I have to believe that for some students this issue caused great personal tension and internal struggle.

In the base of the theater, I snap quick, ill-angled pictures in an attempt to capture the feel of the room that hovers above me, and then Anna and Eleanora look to me as if to say, *Is that enough?* Not wishing to overstep the limits of their kindness, I accompany them out of the room. Eleanora, smiling, says something fast to Anna and disappears. I call *"Grazie!"* to her and wave, though she doesn't turn back to look. Anna guides me around the corner, up a tiny elevator, down a hallway, and into a small room: a library.

"I don't know how much this will help you," she says, "because the books are all in Italian, but you are welcome to look." She leaves me for half an hour to browse, and when she returns, she has photocopied an entire chapter from one book, which she hands to me, along with a bright red hardback, published by the university, on the history and restoration of the anatomical theater. "This book exists also in English," she says. "It is possible you could buy it." I tell her that I'd be interested in doing so and ask her how much it would cost.

"Well, we would first have to get permission from the rector," she explains. I am puzzled. The rector is obviously the equivalent of the university president in the United States, but I am completely unclear on what his role would be in my purchase of this book.

Anna, whom I am starting to think of as Dante's Beatrice, leading me through the anatomical underworld, now guides me down a back staircase and into a forbidding, high-ceilinged office with dark walls. At a large desk with a cabinet behind her sits a neatly dressed blond woman. Anna speaks to her at length in Italian, smiles, gestures at me, speaks some more. Eventually the woman rises from her chair, opens the cabinet behind her, and, from a deep space, pulls a copy of the red hardback that appears identical to the one I have seen in the library, except the title of this one is in English.

"Wonderful!" I exclaim, and, having been told that the price of the book would be sixteen euros, I reach into my purse for my wallet. Anna shoots me a strong, fast look. It is the look a mother gives her child as she begins to misbehave in church. Its universal meaning: Stop what you are doing right now, and, for both our sakes, let's hope that the authority figure hasn't seen that move of yours. I slowly withdraw my hand from my bag, and the blond woman leaves the office, with the book under her arm.

Once she is safely beyond hearing range, Anna explains. "The rector has not yet given permission for you to purchase the book. It is important not to presume that he will do so. His secretary has gone in to ask him. She will return in a moment." When she does, it is without the book. She speaks briefly to Anna, and Anna translates for me.

"The rector is occupied at the moment and will need until tomorrow morning to decide whether to grant permission for the book's purchase." Completely baffled, I return to the courtyard and find the way out. As I do, a young man with a wreath around his neck dances with his grandmother and begins to take off his suit jacket. The next morning Anna tells me that the rector has decided to give me a copy of the book, as a gift, which she hands to me, grinning, with her e-mail address tucked inside in case I have any further questions.

Walking through the streets of Padua, I think of how the young men who observed dissections here in the sixteenth century must have felt the same trepidation and awe that I have while looking deeper and deeper into the body of a stranger's corpse. And though they were physically more distanced from the body than we are, they were actually closer to *death*. My lab partners and I use our own hands, our own strength to reach into the body, to feel its cold wetness, to pull apart its layers and cut away its packing. We touch and cut the body and change its shape in a way that previous generations of students did not. But we deal with far fewer of the realities of the corpse. Our cadaver hosts no signs of decay. It harbors no timeline of rot, no trace of earth clinging to the skin, harking back to an abandoned grave.

M y trip to Padua is a sort of pilgrimage, and in that sense I am not at all alone. Because of its Basilica of St. Anthony, Padua has long been considered a holy site for the Catholic faithful. It qualified as such in the days of Vesalius, and the anatomist made good use of Anthony's relics, which lie in the massive basilica. Today the church—and the relics—receive busloads of visitors daily, who wish to pray to Anthony with requests or with gratitude.

The church named after St. Anthony is enormous and incredibly ornate, seemingly made only of marble and gold. Inside, lines of people swirl around the crypt of the saint. I join the line, and we file behind his tomb. At the back side is a pockmarked granite rectangle—one of the only unadorned spaces in the church. As we wind by, one by one, each person raises a hand to trace the stone.

Past his crypt, in another nave, more lavish and ornamented than any other, is the reliquary. Fragments of Anthony's tunic. Bits of his handwriting. And, in bejeweled, gilded glass globes, Anthony's tongue, larynx, and jawbone. Life-size marble cherubs float above them. The ceiling is dazzling gold. The extravagance is much more striking than the black tongue flecked with pink and set on its thick base, pointing upward; the indistinguishable brown furry mass of the "vocal apparatus"; the ghoulish mandible set in a crystal ball and topped by a jeweled crown.

As I leave the basilica and make my way toward the university, I think only of the fact that four hundred years ago Vesalius is said to have stood before those same relics, touched the crypt, and viewed the holy tongue. Local legend purports that Vesalius was a common target of the enforcers of religious law, since his dissections violated the decree of Pope Boniface. Upon hearing of approaching authorities, Vesalius would leave his anatomical study at the University of Padua middissection and run through the narrow streets into the Basilica of St. Anthony. There he would fall to his knees in prayer, confessing the sins of his research, claiming repentance, and begging for forgiveness for his many transgressions. By the time the church officials reached him at the tomb of the saint, Vesalius was unreproachable in the eyes of the church. That he chose to throw himself at the crypt of a saint whose body the church had cut into pieces to display was a detail that was surely not lost on the anatomist.

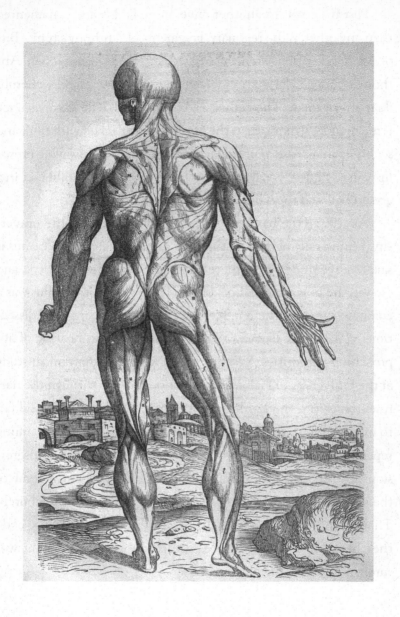

chapter five

Origins of a Corpse

O may I join the choir invisible
Of those immortal dead who live again
In minds made better by their presence: live
In pulses stirr'd to generosity,
In deeds of daring rectitude, in scorn
For miserable aims that end with self,
In thoughts sublime that pierce the night like stars,
And with their mild persistence urge man's search
To vaster issues.

 GEORGE ELIOT, "O MAY I JOIN THE CHOIR INVISIBLE"

I n the semester's third week, the thoracic dissection has ended
and we are ready to begin dissecting the arm. The muscles that
anchor the arm to the body stretch across the shoulder and
onto the back. Therefore, in order to continue with the dissection, we must turn our cadavers over. The bodies are stiff and very

heavy and require someone to hold the legs, another person to hold the head, someone to guide the midsection as it turns, and a fourth person to make sure that the arms turn with the body. Otherwise they would catch and end up twisted beneath the torso's weight. The ankles no longer bend, of course, and so the feet are insistently flexed, preventing the legs from lying flat on the metal table. The neck poses the same problem—the head does not turn, and so when the body is rotated onto its stomach, the face must support the weight of the head. From just such pressures, various parts of the bodies flatten, forever marked by their postmortem positions.

The first cadaver I ever saw was on a tour of a different medical school, when I was applying to programs. The man's body lay face up, but his nose was smashed to one side from the back dissection, which had preceded all others. Somehow that felt like a terrible insult to me, and whenever we turn our cadaver, I hold her face gently in my palm, making sure she rests on her chin rather than her nose, or some other spot that is equally likely to bend. It is illogical. We have at this point removed her heart and lungs from her body, tying them in a brown-black garbage bag and laying them on the shelf below the table for later study. Her rib cage falls to the table as we turn her, and one of her removed breasts lies out to the side of her, facing the ceiling as she lies facedown. Holding the cadaver's chin does little to protect the body's form, but as our actions render her less and less whole, it seems somehow important to preserve whatever human shape we can.

The stiffness of the cadavers comes from rigor mortis, prolonged by the formalin embalming. Several hours after death, the body loses its pliability and assumes a state of contracture, where the muscle tissue is abnormally shortened, preventing any kind of passive stretch. It is the extreme version of the morning stiffness any one of us might feel upon rising. And whereas we can stretch and lengthen our muscles, this action requires energy in the form of adenosine triphosphate (ATP), the powerhouse molecule of the human body. The body works constantly to provide energy to each of its cells. When we die, the production of ATP stops, but a bit of a stockpile remains. So for several hours the dead body can be manipulated and maneuvered. After that, rigor mortis sets in.

Usually, some fifteen to twenty hours postmortem, as the body begins to degrade, the rigor mortis begins to subside. Muscle proteins are destroyed, releasing potent enzymes, which cause cells to burst in a process called lysing. The connection between muscle filaments is gradually lost, and the body regains some pliability. In our cadavers, however, the formalin embalming process cleanses the body of those enzymes, leaving it in a perpetual state of rigidity.

The back portion of the dissection goes quickly for our group. The *Dissector* instructs us to cut one long line through the skin at the body's midline from neck to tailbone. We then make three horizontal cuts: one following the shape of the shoulders, one just beneath the shoulder blades, and one arching across the top of each buttock. The *Dissector*'s illustration looks like a picture from an old cookbook showing the regions of the cow from which various cuts originate.

We peel back the skin and clear away some of the fascia to reveal the superficial muscles of the back. They attach the arms to the central skeleton. The muscles are big and broad, with serious-sounding names: latissimus dorsi and trapezius—large, flat, triangular muscles with extensive ranges across the terrain of the back. We are instructed to "reflect" them, cutting through them at a carefully chosen point and folding the muscles back to see what lies beneath.

The effect is somewhat like opening a triptych, with the muscles swung wide to reveal their undersides and a new layer of musculature, or sometimes bone, beneath. There is always a moment of expectation upon opening, the strange hope of beauty within all that darkness. I think of Fra Angelico painting his famous frescoes beneath the floors of the San Marco Convent in Florence. We see them now, with tricks of light and mirrors, but might they have remained unfound and therefore unseen? Nature is full of dazzling beauty in unexpected recesses. As his lover dies of AIDS, Mark Doty writes about the interior of a crab shell:

> . . . A gull's
> gobbled the center,
>
> leaving this chamber
> . . . open to reveal
>
> a shocking, Giotto blue.
> Though it smells
> of seaweed and ruin . . .
>
> Imagine breathing . . .

surrounded by
the brilliant rinse
of summer's firmament.

What color is
the underside of skin?
Not so bad, to die,

if we could be opened
into this—
if the smallest chambers

of ourselves,
similarly,
revealed some sky.

Compared to dissecting the thorax, the dissection of the superficial back muscles so far seems simple. As we gain more of a sense of how much time is required for each dissection and the study of it, we grow increasingly aware of how limited our time is and how many tasks lie ahead. So we quickly clean and reflect the muscles, find a couple of hidden nerves, and turn the cadaver on her back again. As we flip the pages in our manual to determine the first steps for the arm dissection, we notice that many groups are still fighting their way through thick clumps of fat to reach the back muscles. The slim frame of our cadaver makes dissection less tedious, more clear. My group feels lucky; we can move ahead into the arm while others still have hours of cleaning and separating ahead before they will be done with their backs. We are also lucky from an emotional standpoint.

The table behind ours, with the large male cadaver and his large heart, is getting angry. One of my classmates, Roxanne, is a spitfire of a twenty-two-year-old who has lived her whole life in Rhode Island. She does her dark brown hair in big curls and wears blush and eyeliner to lab every day. She lives with her college boyfriend, who teaches gym to special-ed kids in a small town nearby and earns extra money plowing snow. Roxanne speaks her mind; I like her instantly.

When she sees us turn our cadaver back over to begin the arm dissection, she lets out an exasperated groan, pushes her hair away from her eyes with her wrist, her gloved hands holding forceps and slick with fat, and says in frustration, "Why does our guy have to be such a flippin' horse?" The comment flies in the face of the hushed reverence we have felt the lab required. It is perfect. After a quick moment of silence from both our tables, all eight of us break into giggles and wisecracks. We wait for lightning to strike in response to Roxanne's transgression, and when it doesn't, we all sigh happily into this new place where laughter is both acceptable and sometimes necessary.

We return to our respective tasks with a recalibration of balance. In college a treasured instructor pulled me aside after I had submitted a particularly dark series of poems in a creative-writing seminar. "You know," she said, "Shakespeare was great because he wrote tragedy and comedy with equal vigor." Whether this was advice on writing or a well-meaning attempt to pull a nineteen-year-old back from the edge of writerly angst, the comment resonated then and has continued to resonate for me. Even the most comic moment contains an element of melancholy; even the deepest tragedy harbors a trace of the ironic.

Roxanne's comment then (and the many equally prescient ones that followed) also granted our work in the lab a kind of honesty and perspective. Dissection is work. The hours are long; the toll on our own bodies is felt in both emotional and physical ways. And what each of us knows is that work becomes mundane and demands a sort of levity, whether the profession is that of store clerk or surgeon, counselor or coroner. "I think we can feel free to break for coffee," Roxanne says one afternoon, when frustrated with the lab's slow progress. "It's not like our guy's going anywhere."

Even with the new freedom of humor in class, aspects of dissection remain deeply disconcerting. Beneath each of the dissecting tables in our lab hangs a shiny stainless-steel bucket into which skin, bits of fat, and other unneeded pieces from the body are discarded. The buckets are never emptied over the course of the semester, and we are told to scrupulously place into them anything we remove from the cadaver. The theory is that when the body is cremated at the semester's end, it will be cremated in its entirety. Perhaps the idea comforts potential donors and their families. To us, however, the buckets are haunting. The collection of contents is macabre, and though we try diligently to place all "scraps" in the bucket, we cannot always do so. Hardened blood from the heart is washed down the sink. Bits of skin and fat clinging to dulled scalpels are discarded. I come to think of the bucket as a gesture, and I try to keep as much of the waste in it as I can.

Before beginning medical school, I had heard of students naming their cadavers, often according to some physical trait. The idea of it struck me as disrespectful, and I was sure that I would resist if my lab partners suggested a name for our body. Whether naming cadavers is a means of participating in a medical-school ritual or a human way to make a connection to the body we are coming to know, I am surprised when the name we give our cadaver emerges so organically that I do not resist it. In fact, I embrace it.

When we first remove the damp cloth over her abdomen, the skin is set in firm, deep creases, so we do not notice right away that she has no umbilicus. But for the creases, the skin is smooth, uninterrupted. Already beginning to be trained in questioning morphology, we try in vain to diagnose: "What kinds of things could lead to no belly button?" Silence.

Even if we had known our embryology then, we wouldn't have come up with an answer. *Umbilic*—"the central point." *Placenta*—"Latin for cake." Raj says, "Maybe we got Eve."

O n days when we are all present, seventy-two students huddle over bodies in the wide, white room, and our voices bounce off the painted cinder-block walls. Even at that size, ours is one of the smallest medical-school classes in the country. It is not uncommon for anatomy classes at American universities to take up several rooms, in order to accommodate two hundred or more students, and for faculty instructions to be shown on video monitors mounted in each room. At international schools, where class sizes can number several hundred and cadavers may be in short supply, students may cycle through the lab in rounds. In these arrangements, with sixteen or more students assigned to a single body, the students are essentially only observers of dissections that have been completed largely by staff members.

Sharing one cadaver among four students is a luxury, one to which we have access twenty-four hours a day. Despite spending two entire days each week dissecting with our group members, it is understood that we will not be able to absorb everything we need to learn during that time, so we all routinely make additional trips to the lab. Sometimes the lab harbors a kind of midday clatter past midnight and into the first hours of morning. However, on days with no tests on the horizon, the lab is often quiet, if not empty.

Trip and I occasionally make arrangements to meet one another at the lab in the evenings as the daily lab schedule keeps racing forward. One night we come in to try to get a handle on the coronary arteries, which take the freshly oxygenated blood pumped out of the left ventricle and feed it right back into the muscular heart to sustain its rhythmic beating. The arteries encircle the heart, and as we look for them, we are constantly befuddled. Eve's vessels are small and difficult to locate. Since the body at Roxanne's table has a gigantic heart, we zip up Eve's bag, placing her heart back into the garbage bag that holds her lungs, and head to the table beside us.

Although human anatomy is largely consistent from one body to the next, there is a wide range of what are considered to be variations on the norm. The size of a given structure is one of the most common variables, but shapes and even locations of organs and vessels may differ slightly from person to person and still qualify as unexceptional. For this reason we're encouraged to study the anatomy of many of the cadavers in the lab. Our knowledge base should therefore be broadly applicable, not specific to the one body we've dissected. And yet it is oddly disconcerting to leave Eve and approach the cadaver at Roxanne's table. The familiarity of Eve's body grants us a sort of comfort with the accompanying emotional terrain, and this comfort is absent when we reach beneath the table and pull up the plastic bag, from which we take out the man's heart.

Despite the promising size difference, which we thought would clarify the anatomical structures in the same way the cow's enormous organs did, this heart nonetheless provides its own challenges. As was true for us with Eve, once the heart is removed from the body, it becomes disorienting. We cannot identify one side from the other, and we spend the first few minutes poking our probes into the aorta and vena cava to determine where the channels lead, in the hopes that we can make some sense of right and left, front and back. In addition, the ubiquitous fat on Roxanne's cadaver has not spared his heart. A thick layer of globular pericardial fat encases it, making our search for the coronary vessels all the more difficult.

After a frustrating hour in which we have accomplished little, Trip has to leave to teach an aerobics class. "I have them all doing heaps of push-ups now so that I can tell them about strengthening the pectoralis major and minor!" she chirps, and I grin. I guess we have learned *something*. I walk out with her, having decided to get my anatomical atlas from the trunk of my car and return to try to find the coronary vessels on one more heart.

It's a beautiful, warm September night, and while we walk to our cars, Trip is regaling me with stories of her pet lovebird, Odette, who used to sit on her shoulder while she worked as a hostess at a New York nightclub. Odette has been losing feathers, and Trip tells me that the vet thinks the bird may have a seed allergy. "I mean, can you imagine?" she wails. "I'm supposed to pick the little green ones out of her food!"

When I get back to the entrance of the BioMed Center, it is dark. As I wind my way through the hall, I pass a few students leaving, but as I head downstairs I find myself totally alone.

As we've been instructed, Tripler and I turned out the lights when we left, and so when I walk in to the windowless lab, it is silent and pitch-black. I quickly flip the light switch, and the silence gives way to the buzz of fluorescent bulbs. For the first time, I'm alone in the lab with eighteen dead bodies. I pause in the doorway, take an involuntary deep breath, and hold it. The scene is eerie and disquieting. I feel a little scared and yet know it's silly: the kind of mixed emotion you feel when you hear a noise in your house at night and are *sure* it must be nothing but sheepishly check the rooms and closets all the same.

I tell myself I'm being ridiculous, that things are no different than they were just moments ago when Trip and I were here together. I walk decisively over to Eve, then remember that I meant to look at a different heart. As I stand, trying to decide which table to approach, I hear my pulse, fast and regular, pounding in my ear.

Our lab table is in the corner of the room. I know that it's crazy, but after I've selected a cadaver, I bring its plastic bag of heart and lungs back to my usual seat beside Eve. I cannot concentrate; it is too quiet. I switch on the small radio in the middle of the room which I realize must be there for just such occasions, but I find that my mind, distracted, floats away with the lyrics. I turn the radio off and start to talk to myself about the structures I'm looking for—"So the left coronary artery should be around here, and the circumflex branch should come off of it"—but I feel ridiculous, nervously speaking aloud in a room full of corpses. Before too long it's just too creepy. I tie the bag and replace it, turn off the lights, and head home.

I learn that I am in good company when Dr. Goslow tells me that before he began teaching medical students, he didn't feel as though he had "really 'learned' human anatomy with any confidence or done extensive dissections." As a result, he says, "I lived in the dissection labs for three years. I would dissect into the night, six nights a week, until two or three in the morning. I recall more than once while dissecting alone in the labs that I was freaking myself out for no apparent reason." In order to dissect, we detach from what we are doing, and that detachment is easier to accomplish in a crowd.

For us, in comparison to medical students of centuries past—who dissected unpreserved bodies in various states of decomposition—feeling uneasy in our lab seems foolish. Our cadavers are not only completely formalin-embalmed, but thanks to Arnis Abols, the lab manager who maintains the cadavers—and who for many years used to embalm and preserve them—they are then scrubbed with soap and water. Their hair is cut. Vaseline is rubbed into their hands, feet, and faces. They are covered with fabric drapes soaked in wetting solution. And though they may be stored for some time before they are brought up to the anatomy lab, Arnis changes any soiled drapes and washes the bodies again when they are ready for our use. As is true in our modern funerary customs, we are spared many of the body's unpleasantries, and the elaborate preparation of our bodies enables us to separate from the totality of death more easily.

Yet despite the chemical smell and the distance from the decay of the dead, there are still moments when I am struck by the gravity and the lunacy of dissecting a human body. Andrew, my three-year-old nephew, is getting ready to begin preschool, and to quell his nervousness my brother and sister-in-law tell him that I go to school, too. "What do *you* do at school, Aunt Beanie?" he asks me over the phone before his first day. "Books!" I blurt out. "I get to read lots of books with pictures in them, and you will, too."

Our cadavers are in our lab because they have explicitly chosen to be. Not only has each of them signed a declaration in life that they would like to donate their bodies to medicine, but in our case they have specifically designated that they would like their bodies to be donated to our medical school. Regardless of any of *our* misgivings, they knew what they were signing up for. However, this has certainly not been true historically, and it is not universally true for medical students even today.

During a pathology elective after my third year in medical

school, I spoke with two pathologists, one from Iraq and one from Nigeria, about the sources of their first-year anatomy cadavers. Ade, the Nigerian doctor, described almost exactly the population that might have been found in an American or European anatomy lab decades ago. The dissected cadavers in Nigeria, he said, were all either "homeless," and therefore "unclaimed," or "criminals." He laughed uneasily when he mentioned the criminals, and when I asked him why, he told me that he perceives the government in Nigeria to be incredibly corrupt. Friends arrived home from abroad with money for their families, only to be arrested on fabricated charges and made to pay outrageous fines on the spot to the arresting officers. "So, you see," he said, "the definition of 'criminal' is a little too loose."

Recent papers in the medical literature confirm that Nigerian anatomy departments do not dissect donated or "willed" bodies but instead have access only to executed criminals and the abandoned, indigent dead. In such an arrangement, one paper concludes, "the ethical considerations that attach to unclaimed and bequeathed bodies . . . have implications for the way human material is treated in the dissecting room."

One of the coping strategies most often used by students in the anatomy lab is rationalization, and many of the students quoted in the papers offer arguments that I have either made, or heard from my classmates. "The knowledge to be gained is important for my career," one student says; "[dissection] is an integral part of anatomy in which I must succeed to become a doctor." Another student claims a more altruistic motivation: "I felt it was an activity necessary for the prevention of deaths in the future." Yet another student absolves himself of responsibility by reminding himself that "it is not our making that these people are dead."

And yet the provenance of the cadavers allows rationalization to take on a sinister and vindictive tone. Despite Ade's argument that the definition of criminal was "a little too loose," one Nigerian student took solace in "knowing that the cadavers could not have been good people when they were alive." Another (with the telling use of the words "must have," subconsciously acknowledging less certainty than their absence would) classified the cadavers as "criminals who must have unleashed terror on society and had been abandoned by their relatives." Most hauntingly—and in a way that indicates a deep failure to view dissection as anything but harmful mutilation—one last student asserts that "I considered the cadavers to have been robbers and criminals who deserved no pity."

The Iraqi pathologist, Sam, told me that none of the corpses in his medical school were native Iraqis, but rather that all the bodies appeared to be of Southeast Asian ethnicity. "I have no idea who they were or where they came from," he said. "Maybe Vietnam or Cambodia? But they definitely weren't Middle Eastern." A recent journal paper examining dissection in the Middle East confirms, "A variety of sources have been used to acquire cadavers for medical training ranging from executed convicts to unclaimed or abandoned bodies from public/mental institutions, and also importing cadavers."

A very different model of human dissection occurs in Thailand. The majority of the Thai people are Buddhist, and, due to their belief in reincarnation, there is a cultural reluctance even to donating organs for transplantation. Nonetheless, Thai medical schools benefit from cadavers that are, without exception, voluntarily donated. A recent paper from the *British Medical Journal* explains that the Thai culture deeply respects teachers "to an extent unfamiliar to Westerners," and every Thai donor is granted the highly esteemed status of *ajarn yai,* meaning "great teacher." Two ceremonies establish the status of *ajarn yai:* a dedication ceremony that occurs prior to dissection and a cremation ceremony that occurs at the course's end.

In the dedication ceremony, Buddhist monks and family members of the deceased join students and faculty members in the dissecting room to chant and pray. The names of the donors are read aloud as the title of *ajarn* is conferred, and each cadaver is presented a bouquet of flowers. In a symbolic act of giving to the deceased, meals and gifts are then bestowed upon the monks.

Thai medical students are described as entering into a relationship with their *ajarn yai* that is in the familiar teacher-student model they have known since they were young. The students always refer to the bodies as "great teacher," and never as *sop,* the Thai word for "cadaver." They sometimes bring flowers to the bodies, greet them with traditional bows, and pray for their *ajarn yai* at the temples. In contrast to the American tendency to depersonalize the cadavers with anonymity, covered hands and faces, humor, even rationalization regarding the inanimate nature of the bodies, anatomy classes in Thailand post on each dissecting table the name, age, and cause of death of each donor. Students are expected to know the name of their donor when asked and to retain memory of the name as they practice medicine.

This intimacy yields situations that would be unheard of in Western medical schools. For instance, at the Dalhousie University Faculty of Medicine in Nova Scotia, an introduction is given prior to beginning dissection advising students to notify the faculty if they either know of a recent donor or have recently experienced the death of someone close to them. Any potentially known donor is removed, and, in the case of a recent death, the faculty ensures that the student dissect a body of different sex and age than the deceased relative or friend. In contrast, the *Journal* paper recounts that the grandfather of one Thai medical student had specifically requested that she dissect his body after he died. "She did so," the paper reads, "and was thought to have especially good support from his spirit thereafter."

For each of the donors and the medical students in Thailand, the semester culminates in a ritual procession. Buddhist monks lead the students, who carry the bodies of their dissected teachers to the crematorium. Buddhism itself may help establish this emotional closeness of students with their deceased teachers. Rather than think of themselves, the way a Nigerian student might, as distinct from the bodies they dissect, students in a Buddhist culture may be encouraged to contemplate the fact that their own bodies will eventually be no different from their cadavers.

The Buddhist sutra The Foundation of Mindfulness has in it a section entitled The Nine Cemetery Contemplations in which Buddhist monks are urged to view their own bodies on an inevitable continuum that ends with an unprettied vision of decomposition. "And further," reads an English translation of one of the contemplations, "if a monk sees a body thrown into the charnel ground, and reduced to a skeleton, blood-besmeared and without flesh, held together by tendons, . . . he then applies this perception to his own body thus: 'Verily, also my own body is of the same nature; such it will become and will not escape it.'" The Buddhist teacher Thich Nhat Hanh refers to this sutra as the "meditation on the corpse" and suggests that even lay followers of meditation and mindfulness

> meditate on the decomposition of the body, how the body bloats and turns violet, how it is eaten by worms until only bits of blood and flesh still cling to the bones, meditate up to the point where only white bones remain, which in turn are slowly worn away and turn into dust. Meditate like that, knowing that your own body will undergo the same process. Meditate on the corpse until you are calm and at peace, until your mind and heart are light and tranquil and a smile appears on your face. Thus, by overcoming revulsion and fear, life will be seen as infinitely precious, every second of it worth living.

This type of vision is so entirely avoided by our culture, whose modern death rituals, if they even involve the corpse, center on a body embalmed and made up to prevent any whiff of decay. And so perhaps the question is not whether what is done to the body by doctors and anatomists is wrong but whether the way in which we regard the body—as a "great teacher," as a criminal "who deserved no pity," or as the eventuality of our own physical selves—can be ethically reconciled with the actions we perform upon them.

For me, in the anatomy lab, these questions linger, and their

resolutions are outpaced and temporarily derailed by the revela-
tions our forays into the body offer up. In this way, as in Renaissance
Italy or Buddhist Thailand or Iraqi anatomy labs, the arguments
against the discomfort of medical training are trumped by wonder
and discovery.

A t the end of our third week of dissection, we completely remove the skin from the upper arms of our cadaver, and I feel
some sadness that she is beginning to look less human. Her purple-
brown muscles are now exposed, and the skin on her hands and
forearms has come to resemble formal gloves—like a Magritte
painting where the only thing that belongs is so alone that it appears out of place.

On our course syllabus for the day, Dr. Goslow has written,
"The Brachial Plexus—a nightmare in anatomy?" A complicated network of nerves, called the brachial plexus, stretches from the neck
through the shoulder and the armpit, extending out toward the
upper arm. It gives motion and sensation to the arm and hand.
The cords of nerve begin in the base of the cervical spinal cord
and the uppermost portion of the thoracic spinal cord. From those
nerve roots, trunks of the brachial plexus (about the width of a
piece of yarn, sometimes thicker) emerge, split, braid together, re-
join, split again, and around the underarm divide into the branches
of the three main nerves of the arm: the musculocutaneous, the median, and the ulnar. The plexus is by far the most elaborate structure
we have studied, and due to its placement, and the rigidity of our
cadavers, it is hard to dissect out from the clavicle and underarm.

In the body, nerves look much cleaner than everything else; they seem less constitutionally affected by death. They are white and fibrous, and though we initially struggle to differentiate nerve from artery from vein, experience and texture begin to distinguish one from another.

Arteries are more muscular, as they must often regulate the level of resistance the blood encounters. If you whack your hand against the corner of a table, the arteries expand, or vasodilate, to increase blood flow to the injured area and promote healing—hence the swelling and redness that accompany a wound. To the contrary, if your body is in shock, or is hemorrhaging, the arteries vasoconstrict in an attempt to conserve blood for the brain and heart and other central organs by allowing less blood to flow into the periphery—hence the cold, sometimes blue pallor of a person in physiologic shock.

Veins are less sophisticated, to an extent, yet have a greater capacity to hold large supplies of blood and make it available to the circulation when necessary. It follows logically, then, that the texture of a vein would be less rigid than that of an artery. Think of how much more easily a thin-walled balloon is inflated than a thick-walled one, or how much more effort it takes to stretch a thick rubber band than a thin one. If one major function of the venous system is to distend and one major job of the arterial system is to constrict, then the relative consistencies of their walls make perfect sense.

Nerves, on the other hand, are wholly distinct. Instead of tubular, nerves are fibrous, like a bunch of threads bound together. The nerves feel like thin cords between my fingers, and they are strong. Whereas veins can sometimes accidentally tear away with an inadvertent tug, nerves tend to resist breaking. The trick is in finding them. They are mostly quite thin and often lie buried in fat or fascia. It can take an hour or more of picking fat away from an area the size of your cupped hand to expose one nerve you're looking for. And part of the strangeness of dissection is not even knowing what lies beneath, not knowing wheat from chaff. Early in the semester, we run poor Dale ragged. "Is this something?" we ask at every turn. We don't yet even know for sure vein from nerve from artery from webby fat. "Is this important?" As if there were things in the body that are of no significance whatsoever.

When we find the brachial plexus, we know it is important. Not only are these woven nerves responsible for all movement and sensation in our arms and hands, they are also a huge component of our fast-approaching exam, which is less than two weeks away.

Like the other lab groups, we have (as yet metaphorically) divided the body down the center so that one pair of students dissects the structures on the left side and the other pair performs the matching dissection on the right. The limbs are easily divided in this way, whereas the trunk and organ dissections are necessarily a group effort. Not surprisingly, Raj and I are happiest to be on opposite sides of the body, able to progress with our respective degrees of speed and care. Tripler, who shares the body's left side with me, has begun to loathe time in the lab, however, and sees only lost efficacy in the group's doing two dissections of the same structure.

"I don't understand," she says. "Can't we all four just do one arm *really well* and then spend the time we've saved trying to learn all this stuff that still we haven't had time to learn?" She is paging back through the lab manual; the structures we are supposed to learn are bold and in red print, and the pages are packed with them. "We haven't even learned the coronary circulation yet."

If we measure the learning process by the standards to which we're accustomed from any other course we've ever taken, we should now be wrapping up new work in favor of reviewing the areas of the body we've already exposed. Instead we have two more full days of dissection before the exam, the first of three over the course of the term. In these two days, more than a hundred new muscles, arteries, and nerves will appear, all to be learned by name, function, beginning, and end point. Even certain empty spaces in the body have anatomical importance and their own names—the pulmonary fissure, for example, or the oblique sinus of the pericardial cavity. So the spot at which a muscle splits or the boundary of an organ dips is another landmark to memorize.

During lab, as Raj and I continue the group dissections, Trip and Tamara begin to construct vast charts with columns on the lab blackboards. We try to pool our knowledge and fill the chart without looking in our books and notes. Trip reads the columns to us, and Tamara chalks in the answers once we agree upon them. The board fills. Muscle: *Palmaris longus*, Proximal Attachment: *Medial epicondyle of humerus*, Distal Attachment: *Distal half of flexor retinaculum and palmar aponeurosis*, Innervation: *Median n. (C7 and C8)*, Main Actions: *Flexes hand and tightens palmar aponeurosis*. And another line. And another. Over the course of the next few years, I will say many times to many people, "It's not that med school is difficult conceptually, it's just that there's such an incredible amount of information to learn and attempt to retain." There is no need yet for any kind of original thought. So far our learning is regurgitation at its most pure.

In Italy, amid the creative and scientific energy of the Renaissance, the public viewed anatomical exploration as a new and exciting frontier. However, in other parts of Europe, and in America, the reactions to dissecting cadavers were far less enthusiastic. Even today a natural trepidation exists when we think about bodies as subjects for dissection. Centuries—even decades—ago this unease was compounded by the fact that family members died at home. They were often prepared for burial by loved ones, and their dead bodies were central participants in the mourning process. In addition, widespread belief held that transition into the afterlife would be disrupted if the body could not easily be found in a marked grave. Therefore the absence of the corpse would at least interfere with long-standing funerary customs and important burial rites. At most it would disrupt the eternal fate of the parceled body's soul.

As a result of these misgivings surrounding dissection, bodily donation prior to the 1940s was extremely rare. In the instances when donation did occur, bequests came largely from people who, like the men whose skulls were encased in glass in Padua, had directly observed the need for cadavers for teaching. One entertaining exception is the case of a nineteenth-century British captain who had been denied a war pension. In the hopes of shaming those who had refused him payment, the captain announced his decision to donate his body in the *London Times*, citing an interesting war wound.

The difficulty of acquiring adequate numbers of cadavers—from the time of Vesalius to modern-day anatomy labs in Nigeria and Iraq—has given rise to a constant and ongoing drive to find a substitute for the dead human body. Today three-dimensional digital images are made from living and dead bodies, improvements over the wax sculptures that were used for teaching in the eighteenth century. One collection of such wax sculptures is on display today in Bologna, Italy, which I visited after my stop in Padua. The collection is open to the public but unlikely to be found by tourists—it is housed on an upper floor of a college building in Bologna's Palazzo Poggi, accessed by climbing several unconnected staircases and walking through corridors filled with glass cases of bird skeletons and fish bones.

A marble dissecting table occupies the center of the first room of the palazzo's Anatomical Museum. Along the walls, in glass cases and positioned on pedestals two skeletons stand: a male, holding a staff, and a female, holding a scythe. Their serious props and poses make them look ridiculous. Between the two skeletons, four male wax figures stand in various stages of dissection. Ercole Lelli, the sculptor of the full-body models in the Palazzo Poggi collection, was commissioned to make these models for the University of Bologna by Pope Lambertini in an eighteenth-century attempt by the Catholic Church to dissuade cadaveric dissection.

The irony here, of course, is that despite the intended appearance of the removal of subsequent layers of skin and muscle, the sculptures have never had any such layers. The first appears as if he has been skinned. On the second, the superficial muscles have been "removed" to expose the muscles that lie beneath them. The third is still further dissected, and the fourth displays an empty abdomen, revealing the musculature of the abdominal and pelvic floors. The jaw has now fallen off; with the skull stripped bare, there is no tissue to attach the mandible to it. The waxen upper teeth arch over empty air. The figure resembles Edvard Munch's *Scream*—only a chasm where bone should be.

Interspersed among the pedestaled figures are a hodgepodge of random anatomical constructions of wax and bone. One glass case contains the perineal muscles, which would lie in between the anus and the scrotum or vagina, displayed alongside a model dia - phragm and wax muscles of the hands. In another case, which recalls the relics of St. Anthony, the muscles of the pharynx, larynx, and neck are displayed. Thin straps of muscle are splayed out around the thyroid cartilage so that from afar the case looks as if it contains a pinned-down specimen of rare beetle. Near the window is a glass dome that looks like an African violet globe or a cake plate. In it are a solitary sternum and four ribs.

The second room is more ghoulish. The figures in it were sculpted in the mid-1700s by a sort of wax-anatomy team, Anna Morandi and Giovanni Manzolini. It is hard to pinpoint the difference in these sculptures versus those by Ercole Lelli, other than to say that there is *emotion* in them, a kind of pathos. Several of the works incorporate waxen elements of the recognizably human, which many similar sculptures do not bother with: skin and hair, an intact ear, a seductive posture, a drape of clothing. As in the anatomy lab, when painted fingernails or faces are unveiled, the closer a body seems to life—even in wax—the more disturbing it becomes.

The bones of the skull lie on a navy cloak, alongside a lone open mouth, a severed tongue. The fabric somehow shifts these sculptures not only from clinical to artistic, but also from toneless to emotive. The effect is disarming, in part because of this humanization, and in part because this particular shift away from the scientific introduces a whiff of sensational voyeurism. An open chest cavity is shown, but with intact shoulders and a gracefully upturned neck and chin, just exposing a lovely jawline before a red wax cloth shrouds the implied shape of an absent head, severed from the body at the mouth.

A darkened alcove lies at the end of this room in which the waxen body of a beautiful young woman lies alone. Her head lolls back over the edge of a velvet pillow, exposing her fair throat and the string of gold beads that encircles it. From her neck to the top of her pubic hair, she has been cut open, exposing her opened heart, her great vessels, her diaphragm, liver, gallbladder, kidneys, a stump of sigmoid colon, and also an opened womb, complete with a small, tucked fetus inside. Her eviscerated organs lie at her feet—lungs, omentum, stomach, intestines—and beside her lies a semicircle of the pink-hued flesh of her breasts and abdomen, as if a large puzzle piece or lid were lifted from her living body, neatly revealing her interior. The viewer feels voyeuristic, implicated in a glance that might be akin to that of a serial killer.

As I leave the museum, I spot two glass cases that look like the fortune-telling Gypsy mannequins at an antique arcade. In fact they are waxen busts of the sculpting duo, both done by Anna Morandi. Giovanni Manzolini is depicted as serious, holding an opened heart in his hands. Morandi, on the other hand, has sculpted her self-portrait with a hint of a coy smile. Her hands peel the scalp off of a decapitated head, exposing a bit of brain.

Lifelike as they are, the waxen figures are no more than a three-dimensional textbook—accurate, but completely unable to offer the propelling moments of revelation and functional comprehension of actual dissection. The body is staggeringly complex, and to understand it with any degree of completeness demands dealing with the thing itself—picking up and holding the heart, tracing the path of an artery by threading a pipe cleaner through its lumen. Experienced surgeons return to cadavers constantly throughout their careers, to learn new procedures or practice difficult techniques. The true body as a teaching tool is neither inexpensive nor reusable, but it grants moments of instant and penetrating education: semilunar valves flaring wide and joining with their soft catch of water.

As anatomical exploration across Europe and America began to flourish, the public became more and more fearful of a medical institution that seemed to feed on corpses. Anatomists were not discriminating when it came to the source of their cadavers, and the public started to worry that practically anyone was susceptible to being taken to the medical college and dismembered after death. These fears multiplied when combined with centuries-old rumors that anatomists and surgeons routinely performed vivisections—dissections of the living. Few actual references to such a practice exist, although the *Anatomia Magistri Nicolai Physici*, a manuscript of the court physician in Baghdad at the end of the tenth century, states flatly:

> The anatomists went to the authorities and claimed prisoners condemned to death; they tied their hands and feet, and made incisions first in the animal or major principal members, in order to understand fully the arrangement of the *pia mater* and *dura mater*, and how the nerves arise therefrom. Next they made incisions in the spiritual members in order to learn how the heart is arranged and how the nerves, veins, and arteries are interwoven. Afterward they examined the nutrient organs and finally the genitalia or subordinate principal members. This was the method practised upon living bodies.

Though any acknowledgment of historical vivisection was, and still is, fiercely contested, versions of such gruesome accounts persisted and inspired intense public fear and mistrust of the medical field.

Rumors of vivisection aside, whether it is in fact wrong to dissect the human body is neither a purely historical nor a simple question. Over the course of the semester—indeed over the course of our years of medical training—my peers and I would frequently bump up against feelings that we were doing something innately *wrong*.

As I sit with Eve, my fingers threaded through the nerves of her brachial plexus, following each branch as it joins with and splits from others, becoming renamed at each intersection—radial, ulnar, musculocutaneous—I wonder how I would feel about her had she been a murderess. I know there would be a palpable difference. And had she been impoverished? A ward of the state? A body unclaimed by friends or relatives? A foreigner whose country I did not even know? I wish I could say that those circumstances would make no difference to me, that I would view her just as I do now. I suspect it is not true. The emotional connection that I feel to Eve is due in part to my ability to have sympathy for her lifelessness. In spite of my misgivings about capital punishment, this sympathy would be less readily evoked in me if she were someone whose death was the result of her having killed another.

Eve's body resonates more profoundly with me because I know that she is on the stainless-steel table of her own choosing, and yet that simple decision allows me to make her into the type of person I imagine would choose such a thing: educated, opinionated, concerned with the greater good, unsentimental, rational. Of course, the reality is that some of these things could be true of Eve, or all of them could, or none. But Eve's own agency in her dissection more easily allows me to assign her the traits I imagine a woman of her age would need in order to decide to donate her body to this kind of fate. The fact that the decision was hers, and that it is a special kind of decision that requires a sort of chutzpah, makes it easier for me to think of her as someone like my grandmother. It makes it easier to think of her as someone not so different from the way I see myself.

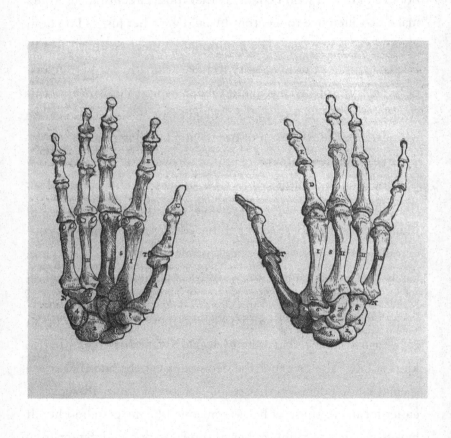

In Pursuit of Wonder

CREON: *About Polyneices:*

>*He is forbidden*
>
>*Any ceremonial whatsoever.*
>
>*No keening, no interment, no observance*
>
>*Of any of the rites. Hereby he is adjudged*
>
>*A carcass for the dogs and birds to feed on.*
>
>*And nobody, let it be understood,*
>
>*Nobody is to treat him otherwise*
>
>*Than as the obscenity he was and is. . . .*

ANTIGONE: *Religion dictates the burial of the dead.*

CREON: *Dictates the same for loyal and disloyal?*

ANTIGONE: *Who knows what loyalty is in the underworld?*

SOPHOCLES, *ANTIGONE*

Once we have finished the upper-arm dissection, we move down the limb to the forearm and the hand. We are a full month into the semester. Raj and I both begin working on the palm dissection, which is difficult in part because the hand feels very human. In order to remove the skin from the hand, we must, by necessity, hold the hand in ours, an intimate and familiar gesture that makes the directive to take a blade to the skin all the more unsettling.

As I cut the palm, I often feel the way one does when you hear about an accident or a gruesome wound. It is as if someone recounted an injury: *The blade went right through her palm— took off the skin.* You would turn your head to the side and wince, suck air through your clenched teeth, identifying with the pain. Because I can call up the feeling of an inadvertent glance of a kitchen knife on the hand much more readily than any cut across the abdomen—and surely because the palm dissection itself requires great pressure and pull—it is an emotionally tiring chore.

The palm dissection is also difficult because the skin of the palm is so much tougher than any other we have encountered. Usually one or two scalpel blades per person last the full seven hours of lab. On the palm alone, I go through five. In lecture, Dale has read us a passage written by Libbie Henrietta Hyman, a 1920s anatomist:

Dissection does not consist of cutting an animal to pieces. Dissection consists in separating the parts of an animal so that they are more clearly visible, leaving the parts as intact as practicable.

In dissecting an animal, very little cutting is required. Cleaning away the connective tissue which binds together and conceals structures is the chief process in dissection. In doing this, use blunt instruments, as the probe, forceps, or fingers. Avoid the use of the scalpel and scissors. You will probably cut something you will need later on. In short, do not cut; separate the parts.

When he has finished reading, he raises his head and says, "That's a warning about your hands. Don't let them look like a weed whacker has been through them."

His voice hangs over us as our patience wears thin. Beneath the skin of the palm is a thick, fibrous sheath, a kind of broad, flat tendon, called an aponeurosis. In lab, Tamara reads us the relevant instructions from the *Dissector*: "Carefully remove the palmar aponeurosis. Do not damage underlying vessels and nerves." Three hours into taking turns dissecting, Tripler and I are ready to hurl the *Dissector* against the wall.

We take breaks from the tedium to work on uncovering the structures of the forearm. Between the muscles of the hand and those of the wrist and lower arm, we are awash in Latin: *Flexor pollicis longus and brevis*, *Flexor digitorum profundus and superficialis*, *Adductor pollicis*, *Extensor carpi ulnaris*. We will learn ten thousand new words in the first year of medical school.

By the end of the day, Trip and I are making dumb jokes that we find hilarious. Dale comes by to check on our progress and offer encouragement, and we tell him we have dubbed the day "nightmarus longus." We think it is the funniest thing we have ever heard.

After the initial frustration of the palm dissection, the following day is completely different. Crazy clarity and order start to arise out of the fat of the palm. Muscles and nerves and arteries are all tangled around each other at first blush, but then, as fat is cleared away, a kind of layout takes shape, along with a larger context.

Two main arteries—the radial and ulnar—provide a supply of oxygenated blood to the hand. The ulnar artery feeds into an arterial arch in the palm, the superficial palmar arch. It looks a little like a traffic roundabout, with branches breaking off a central curve to run up each finger. As we clear away the fascia, we can slide a finger beneath this arch to appreciate its detail. The radial artery meets a similar end at a deeper level, diving beneath the ulnar roundabout to form the deep palmar arch. If you look at the palm of your own hand, there are two main skin creases that cut across it, side to side—a palm reader would designate one as the love line. The crease closest to the base of your fingers is positioned about where the superficial palmar arch lies. About an inch below the lower crease is the region of the deep palmar arch.

The intersection of these two arches and their subsequent shared pathways—and other intersections like them elsewhere in the body—has a beautiful name: anastomosis. The verb form is even prettier—the arches anastomose (from the Greek word meaning to intercommunicate, inosculate; said of blood vessels, sap vessels, rivers, and branches of trees). It is fitting, then, that anastomoses have highly important functionality. They represent multiple possible pathways for blood, so that in case of the interruption of one vessel, as with blood clots, trauma, or atherosclerotic narrowing, the blood's flow in areas of particular importance will be preserved. Not surprisingly, the body's most prominent anastomosis is the circle of Willis, joining the two carotid arteries with the basilar artery to provide blood to the brain.

As if to prove that the body sequesters resources, there is still more fat in the palm of our thin cadaver. We work to clean it away, and beneath the superficial palmar arch are slick, shiny white tendons leading to each finger, seemingly from the center of the palm. As we continue to separate structures from fascia, we realize that for the most part the tendons actually originate in the wrist and forearm and first slide beneath a fibrous band at the wrist (through the famous carpal tunnel) and then again at the base of the palm, beneath the fibrous flexor retinaculum. We discover similar tendons running into each finger on the back side of the hand, coming from the wrist and the back of the forearm.

The sheen of the tendons renders them almost luminous. They look exactly like those in raw chicken—another in a long line of unfortunate links all too easily made between dissection and cooking. They also provide another powerful illustration of anatomic form relating to function. We do not discover this, though, until we ask Dale to come by and help us bring order to the twenty-odd muscles in the forearm alone. Dale reminds us that tendons originate at the end of a muscle and, by definition, connect that muscle to bone. Therefore, since a muscle shortens when it contracts, a tendon necessarily will pull on the muscle and thus tug on the bone in its contracted state.

Dale takes the upturned palm of our cadaver in his hand. "There are two opposing groups of muscles in the forearm," he explains, "the flexors and the extensors. The extensors are on the top of the forearm, and they extend the wrist like this," and here Dale bends back his own hands and splays his fingers as if he were stopping traffic. "The muscles on the underside of the forearm are the flexors," he continues. "They perform the opposite motion," and he snaps his hands down at the wrists and curls his fingers into a fist. "Extension," he stretches his wrists and fingers wide, "and flexion," he curls everything in.

Going back to the cadaver's arm, Dale says, "This makes sense, right? Because if you pull on any one of these tendons going from a muscle on the underside of the forearm to the fingers or the hand, everything bends back toward the spot on which you are tugging." He lifts two different sinewy tendons from Eve's forearm and pulls gently on one and then the other; the cadaver's wrist flexes. "Flexor carpi radialis and flexor carpi ulnaris," he says. One wrist flexor on the radial side and one on the ulnar side. *Flexor* for their actions, *radialis* and *ulnaris* for their locations, and *carpi* for the fact that they attach to the carpal bones of the wrist.

"Let's see if you have a palmaris longus," Dale suggests, and bends my hands back as all three of us look at the inside of my wrists. "Oh, yeah, a really clear one," he says, and he is right. Through my skin, and beneath a few blue traces of vein, we can see a tendon about the width of a shoelace that runs right up the middle of my wrist. "Watch this," he says, and presses hard with his thumb on the tendon, about two inches back from the base of my palm. My fingers curl in. "Not everyone has that tendon, and we don't know why," Dale notes. And then he continues, while I am still staring at the action my body has performed through no will of my own.

He lifts two other tendons in the forearms of our cadaver, and her fingers bend inward even more severely than mine did, as if she were grasping something. "Flexor digitorum superficialis and profundus," says Dale. *Digitorum* for the fingers, *superficialis* and *profundus* for their respective locations at the surface and in the depths of the forearm musculature. "And here, flexor pollicis longus." He tugs, and the thumb comes toward the wrist in a solitary movement. The tendons make a marionette of the skeleton; even in death the bones have no choice but to respond to the strings that pull them.

Once Dale has finished his tour of the muscles, he stands beside our cadaver, as Trip and I retrace the map he has made for us so that he is satisfied that it has sunk in. It has, and we feel relief and some astonishment that order can be superimposed on chaos so quickly by someone who knows the terrain well.

We feel triumphant. In contrast to the endless charts and lists we were compulsively committing to memory, we now understand what it feels like to grasp the function of anatomy.

This moment helps me appreciate Vesalius's commitment to firsthand study of bodies at all costs. As his anatomical discoveries resulting from direct bodily exploration began to resonate throughout the medical world, the authority of Galen's texts slowly eroded. One by one, anatomists aligned themselves with the Vesalian dedication to human dissection, and a profound shift in the foundation of medical education resulted.

Staring from my own curled fingers to those of the dead woman on our steel table, I feel a rush of excitement, a new thrill of definitive understanding. I begin to imagine the energy, the promise that infused young doctors and anatomists when Vesalius's revolutionary ideas began to take hold. Galen's long-inviolable reign of twelve centuries was being toppled by the lowly body, the corporeal, the dead. Yet this burgeoning excitement in no way spread to the populace at large. Who on earth would offer their bodies to such an end?

In the absence of bequests, sixteenth-century surgeons and anatomists began to profit from frequent executions, laying claim to the bodies of criminals. Because the bodies were those of convicted transgressors, their families were in no position to make claims to the right of proper burial. The bodies were taken directly from the scaffold to the anatomical theater, often by surgeons and anatomists themselves, who would come to be seen as the hangman's vultures. Records tell of anatomical theaters and gallows standing adjacent to one another, so that transport could be swift and unimpeded.

Executed bodies were especially prized because of their very recent good health. As opposed to the bodies of people who had died from illness or injury, those of executed criminals were more likely to have been healthy, and they therefore had normal anatomy from which to learn.

In the 1700s the Royal College of Surgeons in Ireland was one of the many sites of anatomical teaching across Europe that benefited from the "subjects" offered by the gallows. However, as a history of the college points out, such an arrangement was not fully satisfactory. For one thing, eighteenth-century students had begun doing their own dissections, and, unlike the cadavers in modern labs, these bodies were not in any way preserved. Thus, instead of two or three cadavers per year supplying an entire cohort of future physicians, two or three cadavers per year now supplied just one anatomy student. Hence, "the demand for subjects greatly exceeded the number made available by the courts of justice." Moreover, the college felt the brunt of the public's disdain, and discretion helped only to a certain degree:

> . . . The founders of the College school thought it necessary to
> provide a rear entrance. . . . The publicity attached to the arrival
> of a criminal's body was not welcomed by anatomical schools.
> The outraged feelings of the relatives and friends of the murderer, whose dissection was for them an intolerable indignity,
> on some occasions led to riot. . . .

Great Britain was one of the first countries to establish anatomical laws, which would inspire similar legislation across Europe and in colonial America. The combination of a growing scientific demand for cadavers and widespread public distress regarding anatomical dissection presented an unusual opportunity for the British government.

In the eighteenth century, sentences for crimes in Britain were severe, and death sentences for relatively minor offenses such as thievery were not uncommon. As the practice of dissecting the numerous victims of hanging went on unofficially for some time, the British government began to recognize that endorsing such a process as official law could have its benefits. Sanctioning the dissection of certain criminals following death would fill two societal needs: Surgeons and anatomists would be supplied with the cadavers to fill the increasing educational and scientific demand, and a judicial sentence even more terrifying than the death penalty could be wielded. The condemnation carried the weighty implication of eternal penance—a sentence not only to end this life but to bring suffering, or at least lack of restful peace, to the life that followed. Hence the 1752 Murder Act was passed, instituting the punishment of "penal dissection" following execution for anyone who was convicted of murder.

Tracing the motivations for and specifications of the Murder Act of 1752 in his wonderful exploration of Renaissance dissection, *The Body Emblazoned*, the scholar Jonathan Sawday writes:

> What was needed, it was felt, was a punishment so draconian, so appalling, that potential criminals would be terrified at the fate which awaited them in the event of their detection. . . . Some new horror was called for which would thwart delinquent desires on the part of the unruly metropolitan populace . . .

Yet although the Murder Act did establish a steady supply of bodies to anatomy theaters, it had the negative effect of reinforcing the public's deep-seated aversion to dissection and an ever greater mistrust of anatomists.

Even with murderers designated for dissection, anatomists still turned to other sources for bodies when the supply from the gallows proved inadequate. Just as for the students of Vesalius, the only alternative source was grave robbery. Across Europe the increasing demand for cadavers gave rise to an entirely new profession: the resurrectionist.

In both Europe and America, profits were to be made for those who could "resurrect" dead bodies from the graveyard and deliver them to the medical colleges. Resurrectioning was an unsavory profession, to be sure, and one that, once its existence was discovered, outraged the public.

Compared to the corpses of executed criminals, bodies that were unearthed were undoubtedly in far worse condition. In all likelihood, mourning proceedings had gone on for several days following the death, so the body would have had as many days to decompose, unlike bodies brought directly from the gallows. Furthermore, the resurrected bodies would have been buried for a variable length of time, adding to the degree of decay. Bodies of the very poor, buried coffinless in mass graves, were often the easiest to disinter. However, they would also have been more exposed to the earth and thus in a more advanced state of rot.

Though "fresh" remains brought the resurrectionist more money, the need of the colleges was great enough to ensure that a use would be found for any subject. In the interest of fetching top dollar, special techniques and tools were developed for exhuming bodies as quickly as possible. Rather than waste the time and energy required to unearth the entire length of the coffin, adept resurrectionists would remove the soil only above the head. A special lever would then be slid beneath the coffin's lid and lifted, to break the lid and create an opening through which the body could be dragged. Next, all clothing was removed from the body and replaced in the coffin. This step was imperative because, ironically, theft of a body would result only in the charge of a misdemeanor, but robbing material goods from a grave was a felony, for which an offender could be hanged.

Many anatomists and medical students could not afford the cost of a "professionally" obtained body and instead resorted to grave robbery themselves. Some medical colleges would even accept all or a portion of a student's tuition in the form of cadavers. Porters in the anatomical theaters were also encouraged to supply bodies and were paid over and above their standard wages to do so. Students and teaching demonstrators would also often supervise the efforts of the resurrectionists. An 1814 letter in the archives of the Royal College of Surgeons in Ireland outlined the requirements for a demonstratorship position. The letter underscored the fact that the professor hired would be expected to "undertake the direction of the resurrection parties."

As more and more graves were disrupted, public outrage increased. The fear of dissection now seized the public psyche. Editorials were written to papers, lawsuits were filed, leaflets were distributed. Expensive, tamperproof coffins were designed, made of metal rather than wood, with spring catches inside to prevent the lid from being pried off; iron vaults were sunk into the ground, and coffins were entombed within them. Some wealthy families even paid armed watchmen to guard their grave sites.

The graves of the poor were inevitably more vulnerable. Poor families took to marking grave sites with flowers or stones or shells, so that any disturbance would be detectable. In a more preventive measure, nineteenth-century African American communities, whose graveyards were disproportionately violated, often mixed the topsoil used to cover the coffins with straw and grasses, in an attempt to make the resurrectionist's spade catch and fail.

In rural Ireland, where graveyards were often set apart from the watchful eyes of town residents, resurrectionists executed less care in their trade, leaving graves open, damaging tombstones and monuments. Cadavers were sometimes even left on the side of the road if the grave robbers feared being caught in the act. Armed relatives of the deceased would often spend every night guarding their family graves until the corpse would have decomposed enough to render it useless—a time period that varied greatly depending on the season. Undeterred, groups of grave robbers began to carry weapons, and battles actually took place among tombstones. Towers were built in which weapons and ammunition were stored, so that relatives could stand watch over the graves from on high.

There are few recorded instances of members of the upper class and policy makers taking up for the poor by protesting the practice of grave robbery. In general such attempts were met with disdain. In the archives of the Royal College of Surgeons in Ireland, the records of such a case can be found. In 1819 in Dublin, an Irish charitable organization proposed funding a watch over one of the city's graveyards for the poor. Professor James Macartney of Trinity College's anatomy department swiftly responded, railing against the proposal by appealing to the pocketbooks and life-styles of the Dublin middle and upper classes. He pointed out that such a watch would harm the financial well-being of the city, since the medical school brought seventy thousand pounds per year to Dublin. Further, he noted, "I do not think the upper and middle classes have understood the effects of their own conduct when they take part in impeding the process of dissection." Specifically citing the practice of the day of using cadaveric teeth to build dentures for the wealthier classes, Macartney added that "very many of the upper ranks carry in their mouths teeth which have been buried in the hospital fields."

Governmental intercession was, however, sometimes a necessity. In Dublin this occurred when dissected human remains were once found floating in the river Liffey. A nineteenth-century county councilman's letter gently mentions this "unsavoury finding" and asks the medical colleges to take greater care in disposing of their used "subjects" in the future.

Anatomists and resurrectionists did not work alone. In addition to employees of the medical school, others with access to dead bodies benefited from the trade. Accounts abound of graveyard watchmen paid to ignore bands of resurrectionists. Undertakers sometimes also played a role, making funerary arrangements with family members and then accepting fees from anatomists to send the bodies to medical colleges instead. Coffins would be filled with sacks of dirt of the body's weight. And, as Professor Macartney pointed out, sometimes cadavers' teeth were removed and sold, as well as their gold fillings.

As communities grew more vigilant in watching over their burial grounds and fending off body snatchers, the demand for bodies nonetheless continued to grow. As a result the resurrection trade became an import-export trade, with cities like Dublin, nearer to rural areas, shipping bodies across the sea to London.

It's not difficult to conjure the unpleasantries that must have been associated with the transportation of decomposing bodies over long distances in a ship's unrefrigerated cargo hold. The inevitable shipping errors also occurred, sending packages of dry goods and foodstuffs to medical colleges and thus presumably delivering the intended cadavers to unfortunate, unsuspecting recipients.

Continued anatomical exploration and discovery fueled the commitment to dissection in Britain, as elsewhere, but the cultural discomfort with such a trend was unabated. By the nineteenth century, it was clear to the British government that the number of cadavers supplied to medical schools by the 1752 Murder Act was insufficient. In addition, the means by which resurrectionists were compensating for the inadequate number of corpses was causing civic unrest. Not only did the routine disruption of graves continue, but eventually—perhaps inevitably, with the least decomposed bodies providing the highest profits for resurrectionists—dead bodies began to arrive at medical schools in "suspiciously fresh" conditions.

The best-known example of this is that of the infamous nineteenth-century Edinburgh duo William Burke and William

Hare. The two Irish immigrants to Scotland did not begin their shady dealings as body snatchers but came upon their trade almost by accident. Hare's wife ran a cheap boardinghouse. When an elderly lodger named Donald died without having paid his past bills, Burke and Hare decided to make up for the loss by selling the old man's body, which, having never been buried, was in comparatively pristine condition. The substantial profit they made off the cadaver, and the ease with which they had acquired it, proved to be more temptation than the two men could withstand. When a second boarder, a miller named Joseph, fell ill, Burke and Hare did not wait patiently for his demise but instead plied the sick man with large amounts of whiskey and then smothered him to death. When that body yielded the pair with a similar amount of money— and a surprising lack of suspicion—their destinies were sealed.

Before the new livelihoods of Burke and Hare were to be discovered, twelve women, two men, and two handicapped children would be murdered and sold to anatomists. Many beggars and street peddlers were among the victims, as well as Joseph, a prostitute and her retarded daughter, and William Burke's cousin through common-law marriage. One youth, known as "Daft Jamie," was a familiar street character in Edinburgh whose absence was noticed. In order to render his body less identifiable, Burke and Hare severed the head and deformed feet from Jamie's corpse. The murderers' tactics were always the same: They preyed on the very poor, promising food and drink and shelter, and always chose someone they were sure to be able to physically overpower when the time came to do so.

Despite the practically still-warm condition in which Burke and Hare's "subjects" arrived at the anatomy school—and the clear evidence of violence that must have been perceived by those who received the body of Daft Jamie—the pair's undoing was ultimately a result not of suspicious anatomists but of neighborly mistrust. In 1828 a woman named Mary Docherty was begging for food and money in Edinburgh, having come there from Donegal, Ireland, to search for her missing son. Burke brought her home with the promise of a meal and invited her to a Halloween party following dinner, where, after she had drunk an appropriate amount, she was summarily smothered and hidden in the bedstraw until Burke and Hare could take her to the anatomy school. In the end Burke betrayed himself, and Hare betrayed him as well. When visitors arrived at Burke's house the following morning, Burke acted strangely and demanded that no guest go near the bedstraw. Curiosity piqued, one eventually did, only to discover Mrs. Docherty's body, now lifeless and cold. Given the chance to spare himself, Hare gave evidence that convicted Burke, who was hanged.

In the ultimate degree of irony the penal codes of 1828 Britain afforded, following his execution Burke was dissected at an event witnessed by select ticket holders. The report of the execution reads in part:

> Early on Wednesday morning, the Town of Edinburgh was filled with an immense crowd of spectators, from all places of the surrounding country, to witness the execution of a Monster, whose crime stands unparalleled in the annals of Scotland. . . . As

soon as the executioner proceeded to do his duty, the cries of "Burke him, Burke him, give him no rope," and many others of a similar complexion, were vociferated in voices loud with indignation. . . . Burke's body is to be dissected, and his Skeleton to be preserved, in order that posterity may keep in remembrance his atrocious crimes.

At the end of the dissection, two thousand students filed past the body, which was then exhibited to the general public and said to have been viewed by thirty to forty thousand onlookers. And even today, in the library of Edinburgh's Royal College of Surgeons, a set of books can be seen whose bound covers were made from the skin of William Burke.

The Burke and Hare case received widespread attention, and the British public seized upon two facts: first, that anatomists in their thirst for subjects to dissect had fueled a market that would devise such a horrible scheme, and, second, that despite the fact that some of the bodies supplied by the murderers had blood crusted around the mouth, nose, or ears, no suspicion was ever reported by the medical school.

These sentiments only contributed to the prevailing mistrust of the medical community and gave rise to a kind of conspiracy theory when, by sheer chance, the arrests of Burke and Hare were followed closely by a cholera epidemic that swept through England. Doctors warned of the risks of contagion and insisted that the sick be quarantined in specific hospitals. The dead were rapidly buried in secluded graveyards. Yet these measures gave scant assurance to the citizenry. Instead the popular opinion was that patients—alive or dead—were not as unwell as the doctors had declared. The consensus was that the sick and dead were being taken in by doctors for experimentation and that all the regulations and recommendations surrounding the disease were a hoax to support anatomists' endeavors.

The traits of cholera as a disease did little to refute these notions. In the throes of the illness, the body's muscles can be gripped by contractions that, after death, suddenly release, giving the perception that some degree of life remains in the body. Similarly, in severe cholera, living patients might turn blue, breathe almost imperceptibly, and feel quite cold to the touch. Such patients might be deemed dead, only to later recover, multiplying fears of being buried alive or, worse, of being vivisected.

It was, in part, this climate of public furor over Burke and Hare, resurrectioning in general, and (however misguided) cholera, combined with pressure from the colleges to provide a greater number of cadavers, that pressed Parliament to look to a new population for a supply of bodies. Their solution was to pass the 1832 Anatomy Act.

Replacing the preexisting Murder Act, this new measure authorized officials to take possession of the bodies of dead paupers. The bodies of the poor would then be transported directly to the medical colleges for dissection. A punishment that had been assigned to the most severe of criminals was now the fate of those whose only crime was poverty.

The act was justified in two primary ways. First, it was cast by legislators as a means by which the poor could repay their indebtedness to the society that had sustained them. If paupers had been given room and board for "free" in homes and workhouses (the work perhaps overlooked as a sufficient form of repayment for food and lodging) and had been unable to pay for it during their lifetimes, then the use of their bodies was a natural—if rather abstract—way for them to negate the debt which had been incurred by their poverty.

Second, by specifying that the act benefited from "unclaimed" bodies, the argument could be made that if there was no one to claim and mourn over the body, then there were no living relations who would suffer from the knowledge that a loved one was being dissected. This was meant to draw a distinction between those thoroughly mourned and carefully buried dead who risked being unearthed and dissected, to the agony of their surviving family members, and those who had no friends and relatives to endure such knowledge. (Interestingly, this may have marked the beginning of a cultural shift in that the main threat of dissection was perceived to be to the feelings of the living, as opposed to the everlasting soul of the deceased.) In fact, the designation between claimed and unclaimed bodies was not altogether clear. It could be construed that a body that was not claimed for burial had no relatives or friends to mourn it. However, it was equally likely that it simply meant that there were no friends or relatives who had the means to pay for a burial, and so the body was left to be buried at the expense of the state.

Either way, medical schools profited from the expansion of sanctioned subjects from murderers alone to murderers and paupers. Certain colleges had more of an advantage than others. If a medical school was affiliated with—or even physically connected to—a hospital for the poor, then it had far easier access to the bodies of the recently deceased poor than did freestanding, unaffiliated schools. Even under such prime circumstances, the mathematics of required cadavers and available bodies did not always square. An 1829 commentary on a small, private medical school connected to one of Dublin's large poorhouses calculated that

> this large pauper asylum does not half supply this small private school, its proprietors being obliged to have recourse to the ordinary means of procuring dead bodies by exhumation. . . . There are . . . at present, in Dublin, upwards of five hundred dissecting pupils; allowing each of them the lowest quantity stated by those examined on the question, that is three subjects each, they would, of course, require 1,500, a number of unclaimed bodies which would, I think, not be supplied by all Dublin, not in one year, but even in ten.

As the commentary suggests, illegal acquisition of bodies continued, even under the Anatomy Act. Ploys to make illicit money from the anatomy trade were spurred on in part by the new legislation, which centered on the claimed versus the unclaimed dead. Though the resurrection trade was predominantly male, women would sometimes be employed to sit at the bedside of a dying stranger in a poorhouse, pretending to be a grief-stricken relative, or to circulate among hospital wards to learn which patients were gravely ill, so that they could show up following a death and have the body released to them. The valuable "fresh" bodies, which might otherwise have gone unclaimed and have been transported to the anatomy labs free of charge, were now sold to the schools from the entrepreneurs who had claimed them. The hospitals and poorhouses had little incentive to thoroughly check the credentials of those who claimed the bodies of the dead, for it meant their employees did not have to spend time arranging transport to the colleges.

Despite governmental intentions, public outrage—and mistrust—over anatomical dissection grew in response to the expansion of the cadaveric pool to include the poor, and to the fact that these legal measures had no effect on the vigor of the resurrection trade. Facts and rumors swirled together to incite an already-volatile climate. The head of a dead three-year-old child was discovered to have been removed from its body and replaced in the coffin with a brick. The head was eventually found in the local apothecary's shop and duly sewn back onto the body. In Aberdeen local dogs unearthed human remains in the backyard of the anatomy school, provoking a riotous crowd to burn the school down. Two medical students were caught in Scotland attempting to steal a body from a grave and, though at first held in a private home, requested to be taken quickly to jail for their own safety. Once the grave-robbing students were imprisoned, hundreds of people surrounded the jail, holding axes above their heads and threatening to kill the offenders.

In his diary from this period, Charles Darwin recounts having seen an angry mob descend upon two resurrectionists in Cambridge: "Two bodysnatchers had been arrested, and whilst being taken to prison had been torn from the constable by a crowd of the roughest men, who dragged them by their legs along the muddy and stony road. They were covered from head to foot with mud, and their faces were bleeding either from having been kicked or from the stones." With language that conflates the resurrectionists' fate with the victims of their crimes, Darwin writes, "They looked like corpses, but the crowd was so dense that I got only a few momentary glimpses of the wretched creatures. . . . [Eventually] the two men were got into prison without being killed."

The 1830s roiled with riots against various and ongoing defamations of the dead, both in America and abroad. The Massachusetts Anatomy Act was passed in 1831 and amended in 1834, similarly allowing unclaimed bodies to be claimed by medical schools. By 1913 the only states that had not passed a law designating the indigent poor for dissection were Alabama, Louisiana, North Carolina, and Tennessee—all southern states whose schools benefited from their large prison systems, which held predominantly African American populations.

For their part, physicians of this period felt betwixt and between. As anatomical knowledge and surgical advancement exploded, the public demanded that medical practitioners have a working knowledge of anatomy and yet continued to protest the cadaveric dissection by which this knowledge was obtained.

The medical information gleaned by anatomists and the resulting advances in treatment were too tantalizing to doctors and policy makers alike to threaten the continued availability of bodies for dissection. Medical understanding, then as now, ignites a thrill in the human spirit. The ability to vanquish disease and heal a failing body combines noble intention with a kind of super -

human power. It can make a doctor look—and sometimes feel—
a little like a god. It can also justify and render ordinary certain actions, like cutting apart a dead woman's body, which would
normally seem impossible to consider.

M y classmates and I are undeniably new to doctoring, but
the lessons that the body teaches us are profound and resonant, and they begin to change us. As our understanding of anatomy becomes more comprehensive, we start to see differently not
only the structure and function of the body but the beauty of it as
well. One day in lecture, Dr. Goslow shows us art slides, and the
entire auditorium slips into a darkened, awestruck hush. On the
broad screen onto which complicated anatomical diagrams and
flowcharts of embryological development are typically projected,
breathtaking images flash: Leonardo's perfect studies of the arms,
legs, hips; a detail of Michelangelo's *Libyan Sibyl* from the ceiling
of the Sistine Chapel, with her massive shoulders lifting the book
of knowledge to the sky. Dr. Goslow shows us pictures of the
hands of Rodin's *Burghers of Calais,* the feet of the scandalously
naked Balzac—both show that odd glory of Rodin, whose stone
feet and hands look more perfectly human than my own, than any
in flesh I have ever seen. In my mind I picture Camille Claudel's
sculpture *L'Âge Mûr,* in which a woman on her knees extends her
arms, reaches desperately toward her aged and hunched love, de-
parting with the figure of death. I see Picasso's *Blue Nude,* seated,
with her back turned modestly toward the viewer. The Elgin Mar-
bles, their flexed quadriceps, their forearms rotated to hold what

must have been a spear, their headless necks. The limbs were what could be understood about the body uncut and unopened. These limbs are signifiers of strength and expression and intent.

We prepare to leave the lecture hall for lab to attempt to link the bravado of Balzac's unclothed stance to the blank, openmouthed rigidity of our cadavers; to look closely at their now-motionless limbs and understand what structures lent them movement.

Just as the lights come on and Dr. Goslow is making his final comments about the musculature of the hand, Arnis Abols, the lab manager who maintains the cadavers, pops into the room to ask a quick question. We are all a little in awe of him—Arnis immigrated to America after having lost his hearing in his childhood, but in spite of this he reads lips flawlessly in English. When he enters, his fingers are a flurry of motion as he signs to Dr. Goslow across the room. The room remains silent, and we all watch Arnis's hands: flexion, extension, adduction, abduction, anastomoses, blood vessels, sap vessels, inosculate, intercommunicate.

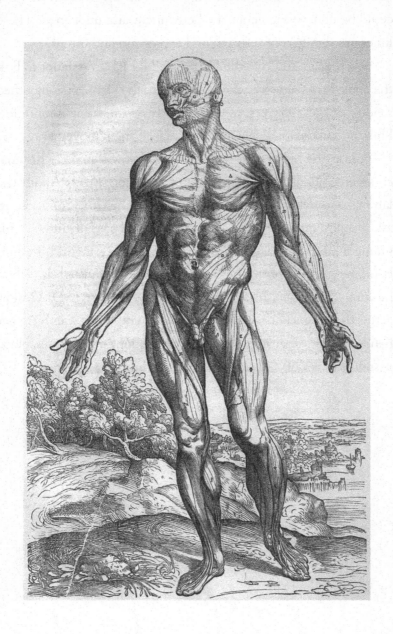

chapter seven

The Bodies of Strangers

Yet why not say what happened?

ROBERT LOWELL

As we are learning the structures in the dead and correlating them to landmarks on our own bodies, we also begin, as first-year medical students and future physicians, to touch the living bodies of people we do not know. And as we continue to wrestle with the question of whether it is wrong to dissect, we find ourselves in positions where we are given more latitude—and more power—than is yet deserved. One of our first-year courses is a weekly medical-interviewing class. We go, in different groups of four, to hospitals throughout the city to learn how to interview patients. Each group has two instructors, called preceptors, with whom we discuss approaches and techniques, as well as the ethics and mythology of medicine. We learn how to ascertain a patient's chief complaint, a one-sentence explanation of why the person came to the hospital. We learn how to write it up using the patient's own language, yet we are also beginning to learn the disturbingly suspicious language of medicine: *The patient reports having had a several-minute period of "crushing chest pain which felt like an elephant sitting on my chest."* The implication is already that what the patients tell us is only what they "report," "admit," or "deny," and not necessarily what we, the relatively omniscient doctors, will find the truth to be.

One of my interviewing preceptors is Rob, a fourth-year medical student in his thirties. He astutely sums up this period of strange transition for us. "You come to medical school like anyone else," he says, "and then within the first week three things happen that differentiate you from everyone else you know. You touch and cut dead bodies. You are able to ask private and socially inappropriate questions of people, and they answer you. And you can suddenly walk into a hospital room, ask someone to take off his clothes, and he will do it." Rob is not flip about this—he is acknowledging our new responsibility and its potential of dangerous power. He also underscores that this phase of learning is an odd time for us, unmarked by any *real* change in our abilities to help people. The patients give us lists of the medications they are taking, and we note them down, having no idea what they do or even how to spell them. They reveal to us their conditions—multiple myeloma, Paget's disease of bone, glomerulonephritis, hydrocele—and we won't learn for a year or longer what those diagnoses mean. It will be longer still before we have any clue how to treat them. Without exception the patients are told that they are participating in teaching interviews and that we do not meet with their physicians or contribute to their treatment plans. Nonetheless, the patients surely tell the stories of their illnesses with the hope that we might unveil some truth in the interview that will alter their prognosis or put to rest their fears.

My friends and family members also ask questions of me in the first semester of medical school. These underscore for me both my utter lack of knowledge and the degree to which an official title like "doctor" or even "medical student" conveys authority. With only a few days of medical school knowledge, I am suddenly fielding phone calls about hysterectomy options and back pain and ear infections. Unless any of my loved ones has a pressing question about how to find their angle of Louis or how to histologically differentiate the upper and lower layers of the skin, I'm of no help whatsoever.

Even in our inadequacy, we are given unprecedented access into people's lives. We wear white coats that instantly mark us as the purveyors of healing and cause dying old men to call my twenty-two-year-old classmates "sir." Family members ask us massive and desperate questions when we enter the room, questions we cannot begin to answer. We ask the patients why they are here; they pull their hospital gowns away from their bodies to show us infected surgical scars across hips, misshapen tumors in their necks and bellies, swollen testicles, amputated limbs. They say, "Do you feel this knot here?" and we put our hands on their joints or skulls or breasts and say, "Yes." I feel your body, your sick and scared body. And I feel how it must be different from how it was, and how it is different from mine. And because I have touched the places to which you direct me on another stranger's form, I begin to know what your neck should feel like, what structures belong. The names arrange themselves in my fingers as I feel from skull to collarbone: occiput, sternocleidomastoid, hyoid, omohyoid, digastric, carotid, jugular. And as I feel them on you, beneath your flexible and warm skin, I know their shapes and their identities because I am picturing the lifeless structures of the body flayed open.

Dissection also provides fundamental lessons that are beyond pure anatomy. In the lab we familiarize ourselves with surgical tools, their proper names, and how they are most effectively held and used. Dale and Dr. Goslow come to our tables and ask us strings of questions about the regions we are dissecting, and in this way we begin to participate in the medical ways of learning that will follow us to grand rounds, the emergency department, the operating room, the outpatient clinic. Perhaps most important are our physical interactions with these human bodies who can neither judge our proficiency nor feel pain from our actions. It is their very inanimacy that allows our natural range of emotions—from frustration to laughter to disgust—to emerge. And so we experiment with the expression of these emotions in a way that we might not be able to do were the bodies live bodies, were the patients true patients.

Just as the cadavers are bodies we cannot harm even as we learn, the clinical medical curriculum has us first practice raising particularly sensitive topics in a controlled environment. One such medical-interviewing topic is acquiring a patient's sexual history.

Each week in the interviewing class, my peers and I take turns interviewing patients, so we have the chance to observe the other members of our small group talking with patients. We are taught to ask "open- to closed-cone questions," which begin with casting one's net wide, then narrowing in to the most pertinent information. We are told to avoid leading questions, to cover the patient's medical history, family history, psychosocial history, to be matter-of-fact and nonjudgmental in obtaining a sexual history and a history of drug and alcohol use. "Can you give me a list of all the street drugs you've ever tried?" rather than "You don't use any illegal drugs, do you?" And "Do you have sex with men or women or both?" instead of presuming that a patient's wife mentioned in his family history is his sole form of sexual contact. Most people are surprisingly forthcoming. A sixty-two-year-old man who has broken his ankle tells me that he has been having erectile dysfunction for the past year when I ask how things are for him sexually. A seventy-seven-year-old woman who is having elective hip replacement tells me that she goes through a fifth of vodka every two days.

When we are first learning to ask about patients' sexual histories, we are duly prepared to respond nonjudgmentally to all types of various arrangements. In order to augment this preparation before we are unleashed on real patients, a group of actors is hired by the medical school to be our "standardized patients." This same tactic of hiring "patients" will be used for various scenarios. As with the sexual-history taking, actors are hired to test our skills of interviewing teenagers. Nonactors are paid to serve as standardized patients for our first physical examinations, rectal and testicular exams, and gynecological exams.

There is much speculation and rumor about who these people are, which perhaps reveals more about our stereotypes than about the hired patients: The gynecological patients are incredibly knowledgeable about the positions of their cervices and ovaries, so some groups report that they are doctors and nurses themselves, altruistically interested in medical education. However, many of the women are large-breasted and quite tan, with severely waxed pubic hair, which leads an opposing rumor to arise that the same women are exotic dancers who do this gig for the money. A preceptor tells us that the middle-aged men upon whom we perform rectal and testicular exams are "well paid" at fifty dollars per student, which leads to a hushed discussion in the waiting room about the life circumstances that would lead a person to accept such a proposal.

The sexual-history taking has none of these physical levels of exposure and so should theoretically also involve less nervousness and pressure. We know that the patients are actors and that their histories are bits of fiction created by the course director, designed to test our ability to access necessary information without making the patient feel uncomfortable. We go to an unused wing of a community hospital, where the actors have donned hospital johnnies and are lying on hospital beds. My patient is a fifty-year-old woman who has recently left a physically abusive heterosexual marriage for a female lover. After the interview I have to stifle a laugh as my preceptor, having apparently forgotten that I, too, have a female partner, lavishly praises my unfazed and accepting reaction to the woman's disclosure of her lesbian relationship.

A shy young classmate of mine named Alan is next, and his patient is a college graduate hospitalized for knee surgery following an injury he sustained in a pickup basketball game. Alan cycles through questions about sexual practice, only to consistently hear from the patient that he has no sexual history of note. The patient has a girlfriend to whom he is faithful, and they are both virgins. They intend to have sex with one another for the first time once they are married. Having been given the set of responses to a sexual-history taking least likely to evoke judgment in a care provider, Alan still manages to render the patient uncomfortable, justifying the hiring of standardized patients in one swift moment. "So wait a minute," he says incredulously. "Just so I understand. You and your girlfriend are both twenty-six years old, and *neither* of you has *ever* had sex?"

Medicine relies upon a balance that is hushed and ill defined. By necessity my classmates and I must have opportunities to practice the skills we will need as doctors. In theory this is no different from any other job: A new waiter takes a small number of tables as a trainee; a young lawyer plays supporting roles in cases and then gradually takes on a few small cases of her own. The idea of inexperience in a medical practitioner, however, strikes us, for obvious reasons, as far more disconcerting. And yet to preserve a constant supply of well-trained and competent doctors, my classmates and I, and the residents we will become, must take on the responsibilities of care for some patients. To become a doctor by merely reading and observing without having managed medications and performed procedures is akin to becoming a licensed driver without ever having been behind the wheel of a moving car. To be sure, our actions are carefully supervised—a hospital pharmacy cannot fill an order in a patient chart written by a medical student unless it has been cosigned by a doctor, and a surgical resident never performs an operation without an attending physician nearby. Nonetheless, the distribution of responsibility—often heavily weighted toward trainees—is rarely made overtly clear to patients, for the precise reason that so few would agree to that arrangement if it was.

In teaching hospitals most minor surgeries are performed by residents and stapled and sutured by medical students. If an uncomplicated procedure such as a hernia repair or a tubal ligation (the surgical term for "having your tubes tied") is scheduled in early July when the new residents have just begun, it is likely that the operating resident will never have done the procedure before in his life. By law, patients must be asked whether medical students and residents may assist experienced doctors in their treatment. However, trainees inevitably do more than "assist," often taking the primary role in patient care.

This balance of information—disclosing enough to be honest without revealing enough to make patients wary—is one routine method of medical training. Another is assigning doctors-to-be to patients who have no choice. Poor and uninsured patients can attend clinics at teaching hospitals and be treated by the residents who are training there. In my third-year obstetrical training, the residents and medical students were assigned to deliver the babies of all clinic patients. The babies that I delivered (while "assisting" residents and midwives) were born to women neither the residents nor I had met before our arrival in the delivery room. In contrast to the women with private health insurance, these patients were all poor and were almost universally some combination of single, nonwhite, and non-English-speaking. I saw no serious complications that resulted from less-experienced care providers. Nonetheless, I knew that when a first-year resident forgot to control the baby's head on delivery, allowing the mother's vagina to tear through to her rectum, or when I knelt between a woman's stirrupped legs to clumsily repair her episiotomy, having only before practiced on a square of sponge, that our training was happening with some cost to those who could not afford to refuse it.

Where, then, is the balance between necessary training and ethics? It is easy for me, in a culture of voluntary donation, to feel squeamish—even self-righteous—about the sources of cadavers provided to Ade and Sam, the Nigerian and Iraqi pathologists, and others like them around the world and throughout history. And yet does the practice of dissecting the unwilling dead differ so distinctly from novice trainees operating on patients who believe that their procedures will be done by experienced veterans? In fact, as the Michigan poet and funeral director Thomas Lynch flatly points out in his wonderful essay "The Undertaking," "The dead don't care." If that is so—as it surely must be, though we are all reticent to acknowledge a truth with such fearful implications for our own selves—then are the unconsented acts that we—indeed that I—perform on the sick and scared and living more problematic than anything Vesalius or students in Nigeria could do to the carefree dead?

This question will be underscored for me one day during my third year when, after having witnessed nearly a dozen routine and minimally invasive vaginal hysterectomies, the procedure was conducted in a way I had never seen. The surgeon enlarged the opening of the vagina by using a tool that resembled an instrument of torture. A stiff plastic circle was anchored between the patient's legs, which were already up in gynecologic stirrups. Then the inner labia were pierced, top and bottom of each side, with small metal hooks. The hooks were attached to thin rubber tubes, and these tubes were stretched taut and pulled through notches around the perimeter of the plastic circle—two and four o'clock on the right side, eight and ten on the left—until they locked in place. The result was like an animal skin being pulled around the frame of a drum.

It feels important to say this here: All surgical procedures seem barbaric to a certain extent. Just as is true with cadaveric dissection, anytime you cut the body, there is an aspect of brutality. For me, I quickly would become able to accept even the most horrible-appearing surgical procedures—cranking someone's ribs apart in order to be able to operate on the lung, punching a circumference of staples around someone's anus to resolve severe hemorrhoids, amputating limbs with necrotic gangrene—once I could rationalize that the procedure was, in the long run, beneficial. Almost all of them would be clearly so. My discomfort would be difficult to quell, however, in any instance where I had seen the procedure done in different ways, and one method—chosen out of habit by the surgeon, or because it required less surgical effort or skill, or because it used a new tool the surgeon was eager to try—seemed to me to inflict unnecessary damage to the patient in the process compared to other methods that were widely available. While the surgeon in this hysterectomy proceeded with his extreme approach, the availability of other less-brutal techniques made it difficult for me to accept his choice.

The first real surgery I see in medical school is live, but on a video screen. Near the end of the first semester of medical school, our anatomy class will go to the nearby women's hospital to see a gynecological procedure. It is not only an introduction to surgery but, unintentionally, it is an introduction to the gray and complex world of medical ethics and decision making.

The doctors and patient will be on live monitors, displayed in the conference room where we sit. There are chairs set up in the room, but we are all so eager for a glimpse into "real medicine," rather than the stuff of our histology and physiology textbooks, rather than our dead bodies, that we sit on the floor in front of the chairs and crowd the television monitors. There are microphones in our room, so that we can ask questions during the procedure, and also in the OR, so that the surgeons can respond or can narrate the course of their action. To begin, the doctors assure us that they routinely acquire consent from patients to have surgeries filmed and shown. But as we are told the patient's medical history, we learn that she is severely mentally disabled after having snorted cocaine years before, which elicited a heart attack and sent a blood clot to her brain, causing brain damage. The scheduled procedure is a hysterectomy due to the patient's cervical cancer, a separate process entirely unrelated to her limited mental capacities. However, her brain damage has meant that the patient was unable to consent personally to the filming of the procedure, and it is her medical guardian who has done so.

The surgeon opens the patient's abdomen, and already we are transfixed. He draws his scalpel quickly and confidently across her belly, and unlike the results of our meeker, more measured cuts on our cadavers, a bright seep of deep red follows an instant behind the blade. A second pair of hands shares the operating space and dabs the metal tip of an electrical cauterizing instrument at the cut ends of the small, bleeding vessels. The current, meeting the wet blood and the fluid-infused flesh, sizzles, lets forth a slender wisp of smoke, and burns the vessels closed.

This technique is used only on the smallest vessels. Larger "bleeders" must be tied off with surgical knots and suturing thread. The ability to stop bleeding with certainty is perhaps the most fundamental priority in surgery; a failure to do so can result in disaster. Two years later I would observe one abdominal surgery in which the bleeding vessel could not be found, and the patient's abdomen filled with blood faster than any of us could suction it. Two of the hospital's best surgeons were called into the operating room while the original surgeon and resident applied firm pressure with their hands to various regions of the patient's interior until they found positions that stopped the bleeding. Only when the other surgeons were in the room, scrubbed, and gloved, were the hands removed and the vessel eventually located and tied closed. The patient, meanwhile, received an enormous amount of blood but would otherwise recover, never fully realizing the extent of the risk she escaped.

Though we cannot tell from our video screen, the smell of surgical cautery is one we will know intimately and one that can only, truthfully if gracelessly, be described as burning flesh. In some surgeries where the use of cautery is substantial, our nonessential student roles will include holding the vacuum suction wand over the surgical field, alternately sucking blood away from the area on which the surgeon is operating and sucking smoke away from the tip of the cauterizer in a minimally successful attempt to spare everyone involved the smell.

Before our televised surgeon begins the hysterectomy proper, he reaches into the new opening he has made in the patient's body and examines the other organs within his reach. He palpates the liver and stomach, ensuring that he does not feel cancerous growths. He pulls her intestines, length by length, through his hands, and when he reaches her sigmoid colon, the S-shaped segment just before the rectum, he finds a nodule that, once biopsied and sent to pathology, is confirmed to be cancerous. The patient's cervical cancer has in all likelihood metastasized, and that bad news, combined with the patient's limited mental abilities, creates an ethical dilemma.

The patient is on the operating table, sheathed entirely in the bright blue disposable paper cloths of the OR. The surgeon pauses, then begins to explain to us the position he, and the patient, are in. In a mentally able person, the surgeon says that he would sew up the patient without performing the hysterectomy, because the cancer has moved beyond just her uterus. In order to attempt to combat the spreading cancer, the patient would then begin radiation treatment of the abdomen, five days a week for five weeks, and then continue further radiation treatments through the vagina. The concern for this particular patient is that she will not tolerate lying still alone in a room for the radiation, much less a vaginal delivery of it. She does not have the capacity to understand what is happening to her, and the surgeon states that, in his experience with this patient during previous examinations, she is unable to be still in the absence of this understanding. The alternative to radiation therapy, however, is grim: Remove the uterus and ovaries in an attempt to slow the cancer, but with less than a 50 percent chance of survival for five years.

The surgeon has an OR nurse dial two of his colleagues from the phone on the operating room wall. He carefully backs away from the table, elbows tucked into his waist and bent, his reddened, gloved hands held aloft, palms facing his chest as if holding a large book. It is a position like a supplicant's. The nurse holds the phone receiver up to his ear, and he speaks quietly, away from the microphones, and we wait, unable to hear.

When he returns to the table, he tells us that he has decided to remove only the ovaries. He does this quickly and sews the patient back up. It feels like a well-intentioned but meaningless gesture—a friendly wave from shore to a drowning man. No option seems to be a good one. The surgeon leaves the room, the cameras are switched off, and we leave the auditorium with our initial excitement changed into something sadder, something profoundly less satisfying.

I walk out of the hospital that evening and make my way through the enormous parking lot to my car. Behind me the huge complex of bricked medical buildings hovers, windows lit, smokestacks releasing great plumes of steam into the sky. The buildings are tall and sprawling, consisting of three hospitals in total, with underground tunnels and aboveground passageways connecting them for the transfer of patients in beds and wheelchairs from one to another, for employees to keep warm during the bitter cold of New England winters. The years have seen a number of extensions and expansions to the main buildings, giving the space the kind of eraless but dated look that most hospitals have. Glancing over my shoulder at the squares of light leaking through institutional rose-colored curtains, I think, *There are sick people in all those rooms*. I start to count, then try to multiply the number of windows I can see times the number of floors I can see times the two or three or four people per room. The numbers swirl, mix with physiology formulas: *The diameter of a vessel times the velocity of blood flow times four patients in a room, all demented, with a twenty-year-old nurse's aide watching them. Equals hopelessness, equals years of sickness and a room that smells like feces, equals wrists tied to bed frames with restraints, equals a young woman with no mind left who cannot understand radiation through her vagina. Equals death. Or something worse.*

The planted trees in the midst of all the pavement are already almost totally bare, with only a few stiff leaves, curled and defiant, holding on to the slim branches. I haven't remembered to bring gloves, so, once I'm in my car, my fingers are freezing on the cold steering wheel. I drive down the city's historic street; the streetlamps are flickering yellow, and the Federalist houses with their stern, flat fronts appear to be standing guard. For what? Their eighteenth-century landowners? The lamplight? The cancer patient with the uterus she was meant to have removed? Me? All of us? Water appears on my windshield, too substantial for mist, too fine for rain.

When I unlock the door to our home, our dog, Maggie, scrambles over from her space on the rug's edge, fights and fights the nearly irrepressible urge to jump on me, runs in circles, whimpers. Louis Armstrong is playing, and Deborah is on the fat green velour couch, pencil tucked behind her ear, student papers strewn around her, a half-drunk mug of tea at her feet. She looks up, grins, waves bleakly at the papers as if to say, *Terrible, all of them*, and then asks me how the surgery was. From the stereo on the bookshelf, a Cole Porter song: *A trip to the moon on gossamer wings / Just one of those things*.

I tell her two things, both truths, the first comfortably removed and political. I tell her that I learned that hysterectomies remove most or all the lubricating capabilities of the vagina and that some result in vaginal shortening. I tell her that the hysterectomy is the most commonly performed surgical procedure in America, and we launch into a long, lighthearted discussion about how if the most common surgical procedure was one that resulted in erectile dysfunction and penile shortening, there would certainly be a great bloom of innovation to find alternatives.

The second thing I tell her is that I am stunned by how roughly the body is treated during surgery, how the intestines were pulled out of the body and shoved back in, then mashed on one side, literally tucked tightly into the body beneath a towel and held back by huge metal clamps. I tell her that it is not at all the way one would intuitively treat a living body. She is quiet, then hypothesizes that maybe this is the purpose of doing the uncomfortable, forceful procedures in anatomy lab. I nod.

I don't tell her about my ambivalence about the consent, about the metastatic cancer, about the hopelessness of the options available, about the futility of the surgery. What training must I go through in order to prepare for those things? If she is right, if to be comfortable cutting open an abdomen to look for disease, I must first saw open the skull of a dead woman and remove her brain, then what preparation will I have for the more awful moment when there is no action in my arsenal that can provide even the smallest bit of relief, much less any sort of cure? *One of those bells that now and then rings, / Just one of those things.*

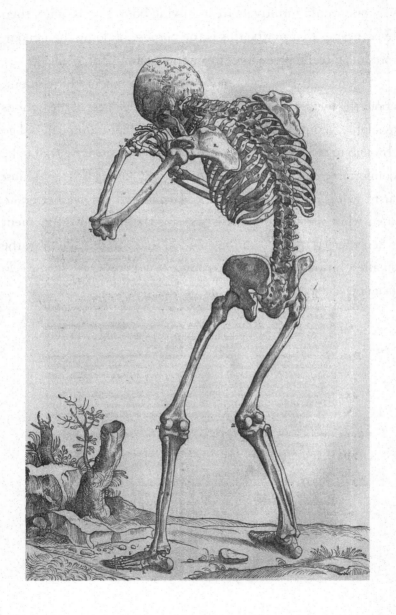

Toll

Medical training is more accurately characterized as a process
of resocialization, that medical students are not only "passing through" an
experience intended to refine previously established values, but that those
previous identities must be repudiated and new and countervailing
identities assumed.... Medical students are not so much confronted
with an absence of knowledge as with the collision of simultaneous and
conflicting values, norms, attitudes, motives, and emotions.

GEORGE E. DICKINSON, ET AL., "DETACHED CONCERN AND
DEATH ANXIETY OF FIRST-YEAR MEDICAL STUDENTS:
BEFORE AND AFTER THE GROSS ANATOMY COURSE"

Most nights, in the dream space between wakefulness
and sleep, I am skinning people.

Not that I can see myself standing at the table in
lab. It is just my hands, and the left pulls back skin from some un -
identified part while the right fingers sweep away the fascia beneath.

It does not feel depraved. It is the slight annoyance I felt as a breakfast waitress when, in the same near-sleep moments, I would think, *Syrup to table twelve. Extra butter to twenty-six.* As then, I wake myself with logic—there is nothing to do now but sleep. Nothing to do here but rest. Then eyes closed, deep breaths, and the returning pull of the left, sweep of the right.

Wake, logic, breathe, left pull, right sweep, wake.

In the deeper moments of sleep, the body is in bed with me. I roll up against her. Feel her weight on the mattress. In night paralysis I cannot lift the body away. I beg her to get up, knowing she can neither leave nor hear my request to do so.

I wish her gone.

By the end of the first month of medical school, my classmates and I become further immersed in our dissections, and we begin to compartmentalize our responses to our cadavers. We are more and more comfortable in lab. The quiet of our initial forays has given way to a loud buzz; in the lab's break room, we pull a skeleton over and drink coffee and snack on cookies while quizzing one another on muscular insertions and names of bony prominences. I hold my mug in one hand and use the other to lift the skeletal arm or hand, to wheel the fleshless body around so its back faces us while I touch ridges and joints. We are increasingly able to detach from the strangeness of our actions, which serves us well. The academic pressures not only in anatomy but also in our other courses are mounting, and the stress of the incessant workload and looming exams demands an increasingly great share of our emotional reserves.

With the approach of our first anatomy exam, the tension in class rises dramatically. We still have new material in the arm to learn, we have memorization left over from the thorax to finish, and we have to review the shoulders and upper back, which have been all but forgotten. We feel as if we were bumping up against the upper limits of our memories. Newly learned structures seem to directly displace the memory space of those that were learned just two weeks ago. Our first lab practical is mere days away; we have almost no energy for deep mortal reflection.

Still, the suppression of this necessary contemplation does not come without a cost, and we are reminded of it when we leave the business of lab, in our vivid dream lives, when the stress lightens for a moment after an exam and instead of relief we feel an uneasiness we cannot explain away.

The night before our first anatomy exam, a surprising benefit of the anxiety we feel is that our discomfort is completely supplanted by stress. We will have both a traditional written exam and a lab practical, in which we are asked questions directly pertaining to the cadavers that are positioned in front of us. The practical, in particular, has the reputation for being grueling, not because the instructors intentionally craft a difficult exam but rather because we are by now supposed to be able to identify several hundred structures and know their locations, beginning and end points, and functions. We must know the branching points of veins and arteries, the nerves that supply each muscle, and the type of nerve fibers in each nerve. We must locate and identify structures in photographs, X-rays, CT scans, and cadavers. In addition, for the written exam we must also know the embryological development of the structures we've studied and be able to begin to solve clinical problems.

The volume of information for which we are responsible is absolutely enormous, and my mastery of it is spotty at best. I know the muscles of the hand and arm down cold but falter on their innervations. I know the common symptoms of brachial plexus injuries but have forgotten much of the embryology of the nervous and circulatory systems. As the list of items for which we are responsible grows, I accept that no matter how hard I study, there will be questions on the exam whose answers I simply do not know.

The lab closes on the morning of the day prior to the practical so that the faculty members can arrange the bodies and mark them with questions. To medical students, most of whom thrive on studying up until the very moment testing begins, this lack of access is disorienting. The hours leading up to the lab's closure are frantic, with many students promising to stay through the night until the 8:00 A.M. locking of the doors.

In a show of support for us, the faculty members take shifts at the lab, there to answer questions and to explain the format of the exam. A pot of coffee is constantly brewing in the break room, and racks of clean mugs await us. A couple of the instructors have even brought in popcorn and cookies, "to make sure you guys are consuming something other than caffeine," but we know that it is simple kindness and an attempt to make the whole process more bearable. To an extent it works, and the late-night session takes on a casual feel, despite the buzzing undercurrent of anxiety. When I grow tired and leave at 11:30 P.M., I am one of the very first to go and probably the only one heading home, rather than to the science library.

I have concentrated on preparing for the practical, rationalizing that I will have an entire day to study the material for the written test once the lab has closed. By the time I get home from the lab, Deborah is asleep. She has left one lamp on for me in the living room and the closet light on in the bedroom, but the rest of the house is dark. It is past midnight. I am bleary-eyed and know myself well enough to realize that any attempt at serious studying at this point will be unproductive at best, and counterproductive at worst. I know equally well that my classmates, many of whom are younger than I, will surely stay up at least several more hours. I am grateful for the excuse of five or six extra years and give only a cursory glance to my notes from the evening before taking off my formalin-steeped clothes and stepping into the shower's stream. As I try to rub away the smell, I am barely conscious of the recitation in my mind that follows the bar of soap across my shoulders and down my arms: *deltoid, supraspinatus, triceps, biceps, brachialis, flexors, extensors*. I feel the bony suggestions of my ribs and clavicle and scapula, my humerus and radius and ulna. The hot water makes my heart beat fast, and I picture the blood on its course through my lungs, to my heart, out the great vessels and the arteries of my arms to my fingertips, which will still smell like the lab when I climb into bed, despite all the scrubbing, all the soapy lather and shampoo.

The next day goes quickly. I begrudgingly attend my histology and physiology lectures in the morning, but my mind is elsewhere. The few of us who are in the lecture hall absentmindedly take notes while flipping through charts and index cards covered with muscular origins and insertions, the branches of the brachial plexus, thoracic innervations, embryologic derivations. Everyone looks a bit disheveled. A few students are in their pajamas. Baseball caps abound. One woman winds around her wrist what appears to be a baby blanket. As soon as the lecture ends, we rush out, saying little to one another, exchanging feeble and sympathetic smiles.

I head straight home and eat a fast lunch before spreading out my charts and notes to study. Unlike the lab practical, the written exam will cover material that is more conceptual, rather than identification-based. I draw and redraw embryological stages of development and memorize embryonic structures and their adult correlates. I read through various clinical situations and try to determine what type of injury would have led to the patient's symptoms. The phone rings, and I turn off the ringer rather than pick it up. The mail is tossed through the slot in our front door and lies scattered in the entryway for hours.

At some point I stand up and stretch. It is an unseasonably warm day, and I look out the window to the house across the street, where two shirtless construction workers hold a ladder. They pull a rope, and the ladder extends. The one: legs extended, arms extended, shoulders flexed to sturdy the ladder and grasp it with flexed fingers. The other: left leg extended, right leg flexed at the hip, flexed at the knee, foot dorsiflexed and resting on the ladder's second rung. Right arm extending overhead to grab the rope, then flexing at the shoulder and elbow to pull the rope in and raise the ladder up; the left does the same: extend, flex, pull, raise. That is what I know now. They grimace, wipe their damp foreheads with their gloves. The tarp billows loud blue, distorts the shadow of the ladder into uneven rungs, a strange diagonal. And yes, their muscles change shape, each one acting independently as if contributing to the whole concerted motion in spite of itself. The man highest up on the roof raises his right arm to hammer. His shoulder contracts and swells, takes up a little more of the sky.

The afternoon flies past, and by evening I have not yet made it through half my list of topics to review. Quietly, Deborah makes dinner and sets it in front of me, then, sometime later, takes the empty dishes away. Medical training can sometimes demand so much space and time as to prevent even the most basic of extracurriculars. Long lecture and lab hours dominate the first two years, with frequent nights consumed by study. The hours get worse, and working a dizzying eighty or more hours per week during the clinical years of medical school and residency means that you're never even outside the hospital during times when banks or post offices are open. If you don't have stamps before you begin a difficult rotation, bills do not get paid on time. Your own medical appointments are impossible, both to schedule and to go to. I grow my hair long so that I can go months without a haircut. When my best friend asks me to be the maid of honor in her wedding and to wear a tea-length black dress, I order five dresses in the middle of the night off the Internet, repackaging all but one to return someday when I can make it to the post office during business hours.

The sheer number of hours my classmates and I devote to our medical education is staggering, but significantly fewer than students faced before recent media attention on sleep deprivation and medical error forced some regulation of trainee work hours. The first two years of school require the absorption of tens of thousands of pages of textbook material. Those who hope to go into the highly competitive fields, such as dermatology or ear, nose, and throat surgery, may see their chances severely compromised by just a few low test scores, even at this early stage. I am relatively lucky as far as the pressures go. My field of choice, psychiatry, is not as brutally competitive. When I see the surgeons-to-be frantically studying moments before exams, or their crushed expressions after receiving scores that are in all likelihood higher than a score with which I am well satisfied, I'm reminded that my own all-encompassing experience of medical school is a good deal less pressured than it could be.

Still, I lose that frame of reference right before an exam, when my mind is hijacked by the material I must absorb. Even though I'm exhausted when I climb into bed, my mind refuses to quiet. When I do sleep, it is fitfully. One night I dream that I am given a teaspoon and told that I have seven hours to swallow every spoonful of the sea.

Meanwhile the workload in our other courses is mounting as well. We spend off hours in dark rooms peering through microscopes at pink-stained slides of tissue trying to discern the difference between the cells of trabecular and laminar bone, between duodenum and jejunum, between esophagus and vagina. We find empty lecture halls and use their blackboards to chart how fear leads the nervous system to increase a body's heart rate. Feelings of academic and emotional inadequacy creep up on all of us.

The worst moments for me are when we meet in groups to work through physiology problem sets. I do not have a natural mind for numbers, and when we are asked to calculate voltages within the heart, ion concentrations in muscle cells, and fluid levels within the body in various states, I cringe. The poetic names and visible structures of anatomy and histology are pleasing to me, but I can grasp the chemical abbreviations, numbers, and symbols we have to be able to manipulate only, as Lex would say, "by beating it through my thick skull."

At the start of the year, in his opening speech, an associate dean warned us that we would each feel ill equipped at times over the course of these years, that we would wonder whether we are up to the challenge, whether we belong. But when these moments of intellectual disorientation and feelings of inadequacy overtake us, we are nonetheless unprepared. We *know* we are fraudulent. We *see* that our peers are not struggling in the same ways. Other students ask us to explain things to them, which we do. But we think that those are the easy things, the concepts they would figure out in another second on their own. I think, *It is this really tough stuff, which they are breezing through, that I simply cannot comprehend.* I blankly nod along during another physiology problem set. "Oh, yes, the action potential would occur here, so the ions would flow outward, I agree with all of you."

Perhaps cowardice keeps me silent, not wanting the other group members to know that I haven't the slightest clue how they arrived at their conclusion, but there is also a degree of self-sufficiency mixed with self-reproach. *I can figure this crap out*, I say to myself. *I just need to spend more time with it.* So I keep quiet; no purpose in slowing down those of my peers who get it. I'll figure it out on my own. And I try to do so, though with varying degrees of success.

Two years later, in my third year, a group of first-year students asks me to lead a group with one of the university chaplains that would help medical students openly acknowledge and deal with the personal burdens of medical school. I am taken aback when so many versions of my own experience surface. A tired man in his young twenties, wearing a too-big sweater and jeans, speaks softly, as his eyes well up and threaten to spill over. "You guys don't get it. I mean, I *really* don't know why I'm here. I don't even understand some of the questions that get asked in lecture, let alone the answers." One man breaks down, and the jig is up. The emperor is buck naked. The other first years climb all over each other to say, *Me, too, I feel the same way. Did you know that the woman who asked that question about neuroanatomy yesterday did graduate work in neuroscience before she came here? I can't believe you are saying this; you're the one who explained the hepatic portal system to me over gyros when I didn't have a clue.* We all look to him, think of our individual moments that felt exactly like his, breathe deeply and say, *Wait a minute; that's me.*

The warm and affable chaplain provides startling perspective from another world. "Do you understand," she says, "how crazy it is to say, 'I am not working hard enough' when you just said that you have not spoken to any of your friends in weeks? How can you ask," she bolsters them, "'Can I be trusted to make the right decisions for my patients?' when you have spent the last two hours explaining how you felt such a strong desire to help a sick old man you were only asked to interview?"

We understand that it's vital to gather for this discussion group outside the grounds of the medical school. Anywhere near the labs and lecture halls, the tension is inescapable. Trip and Lex and I learn early on that staying in the BioMed Center to study, or going to the science library, is a waste, not because the atmosphere is unproductive but just the opposite. We are all comparing ourselves to one another, and doing so while studying can be deadly. Trying to concentrate on a histology chapter is impossible when a nearby group is efficiently racing through anatomical charts of innervations you haven't even begun to look at. Conversely, staying calm when your friend at the study carrel beside you is a bundle of nerves just doesn't work. I study at home on our couch, notes and open textbooks strewn across the living-room floor, interrupted by half-empty mugs of tea and, often, our cocker spaniel curled up on a spread of cards. When we think we would benefit from group study, Lex, Tripler, and I meet at one another's houses, brew tea, and eat muffins. We have learned that there is something unbearably toxic about being in the medical-school building outside class time.

The tension of trying to memorize and understand so much information is compounded by the fact that many of us know we are suppressing a good deal of emotional turmoil about going back to lab several times a week to dissect a human body.

The relative dearth of references in the medical teaching literature to the subject of psychological coping and emotional preparation for human dissection speaks to a long history of stoicism in medicine. Cultural historian Ruth Richardson writes of the anatomy classes of the nineteenth century that:

> no warnings were offered the novice of the carnage an early nineteenth-century dissecting room would reveal—no indication of the need mentally to prepare for a first experience of the practice of dissection, nor the need to control repugnance, fear, nausea. Manuals of dissection seem studiously to have avoided reference to the potentially unpleasant nature of the activity they describe, and its moral or philosophical implications. The language used epitomises "detachment" and emotionless "objectivity." Reference seems entirely lacking as to the means by which bodies were obtained, to the identity or personality of the dead person upon whom the student might spend many hours toil. Indeed, any notion of the individuality, even the *humanity* of the body was lacking, as also was reference to the theological or teleological significance of its destruction.

The only real tradition of emotional preparation seems to be that of the telling of cadaver legends. The field of anatomy is riddled with tales that are the anatomical equivalent of the urban myth and have persisted through generations. Like the urban myth, such stories tend to begin with an assertion of reliability in which students report that, for example, they were told the story by a friend in medical school at Penn who witnessed the event or that such things happened routinely "when my dad took anatomy." There is the story of students inserting a severed penis into the vagina of a female cadaver and the one of a group of students said to have given change to a toll-booth worker from the amputated hand of their cadaver. There are tales of Halloween parties decorated by dismembered limbs and ears, others of medical fraternities putting dissected cadavers in sorority-house coat closets.

Medical educators are only now increasingly investigating whether the coping mechanisms utilized by medical students are healthy ones. The value of anatomy legends was explored by George Dickinson in a fascinating paper, "Detached Concern and Death Anxiety of First-Year Medical Students: Before and After the Gross Anatomy Course." Dickinson notes that the message of the horror stories so gleefully told by more senior medical students is clear: The ability to withstand both the experience of dissection and the scandal of cadaver legends without blinking is the first step in taking up the mantle of "doctor," of comfortably straddling the delineation between life and death. Failure to laugh off the tales or to dissect bravely and confidently is symbolic of a deeper inadequacy, of not being up to the job. Dickinson writes that "cadaver stories portray, and thus help create, a world of outsiders and insiders and emotionally weak or tough medical students. A norm within medicine is that one must have emotional strength to survive the mental stress brought on by practicing medicine. Strength begins and must be displayed in gross anatomy."

One tale that, like many of the anatomical legends, is eerie and disconcerting but, unlike the myths, is widely considered to be true, is the story of William Harvey's seminal 1628 study *De moto cordis et sanguinis*. When asked about the origins of this revolutionary work, which revealed the previously unknown circulation of blood within the body, Harvey explained that he had come to the conclusions published in his research "by autopsy on the live and dead, by reason [and] by experiment." This fact might not have been so utterly remarkable had Harvey's autopsies not included his dissection of the corpses of his own father and sister.

"Anatomy is the Basis of Surgery," the famous eighteenth-century surgeon and anatomist William Hunter told his beginning students. "It informs the Head, guides the hand, and familiarizes the heart to a kind of necessary Inhumanity." This "necessary Inhumanity" would be characterized in today's medical language as clinical detachment or detached concern. However named, the concept acknowledges that dissection—even when not performed on the members of one's nuclear family—demands a great deal of emotional suppression.

The reality, of course, is that students—male or female, with or without doctors in the family, Raj or me—*are* affected by the process of human dissection. The degree to which we are affected varies, and gauging our responses is complicated by the powerful social pressures on all medical trainees to conceal our anxieties. Anatomy lab is a place where students must learn to tread carefully in the emotional minefield of medicine. One study found that medical students ranged from reflective to very distressed about their role in human dissection, with roughly half reporting significant emotional upheaval. Publicly, however, all the students rigorously adhered to the norm of stoic silence. A second study conducted at Stanford, by Peter Finkelstein and Lawrence Mathers in 1990, supports this finding. The emotional and psychological responses of Stanford medical students to the human anatomy lab were studied over the course of four years. A large proportion of the students reported some degree of psychological upset, with some students reporting severe disturbances, including "nightmares, intrusive visual images, insomnia, depression, and learning impairments" in response to their experience in the dissection room.

The study compared the symptoms reported by students to those expressed by people who had witnessed or experienced significant trauma, and the overlap was striking. In fact, the degree of similarity in symptoms led the researchers to conclude that the reactions reported by some students showed a close correlation to experiences reported by sufferers of posttraumatic stress disorder (PTSD). The typical image of a person susceptible to PTSD is a soldier who has seen friends brutally killed or who has himself committed wartime atrocities. But survivors of other traumatic events on the extreme periphery of normal human existence, like natural disasters or sexual abuse, may also suffer from PTSD. The symptoms reported by the Stanford students assert that, for some students, cadaveric dissection may be in this same category of traumatic experience.

Many other studies over the years have shown that dissection is seriously unsettling to students. It generates conflict between the natural, emotional response to a disquieting situation and the necessary detachment a medical provider must bring to work. Finkelstein and Mathers's Stanford study goes beyond these findings, however, and actually asserts that medical students' psychological steadiness has the potential to be profoundly disturbed in the anatomy lab. The study concludes that anatomical dissection demands that a personal and psychological transition occur, requiring "that after [the] course is over, [students] arrive at a new state of emotional balance which includes the powerful death images and experiences of dissection." In other words, the emotional equilibrium that Trip, Raj, Tamara, and I held prior to beginning medical school is irrevocably altered to include our visions of Eve and the actions we have performed upon her body.

The first-year med student learns that finding a way to reach this new equilibrium is essential if she wants to be given the keys to the field of medicine. If academic performance in the first semester of medical school is meant to ascertain a student's ability to deal with the breadth and complexity of knowledge required for medical practice, cadaveric dissection is a measure of one's capacity to handle the emotional demands of being a physician.

Certainly the act of dissection is at the periphery of the normal human experience and requires engagement with the outer range of human emotions. Yet the results of the Stanford study were—and continue to be—met with fierce argument. Indeed, the authors point out that their assertions run counter to the approach to anatomical teaching that dominated the twentieth century. "The prevailing view expressed in both course planning and in much sociologic commentary," they write, "is that any potential psychological stress involved in participation in such a course will be negotiated quickly and uneventfully. One senior professor at Stanford, with nearly 50 years' teaching experience, said that in his view students' adjustment to the dissection experience was 'immediate and complete.'"

As much as the professor's statement may seem incomprehensible in light of Finkelstein and Mathers's findings, some students do seem to cope extraordinarily well. In his own study of an anatomy class, the sociologist Frederick Hafferty found two basic groups of students. The first group considered their cadavers to be purely biological specimens, like the cats, frogs, and earthworms dissected in other classes. The second group viewed their cadavers as formerly living human beings. Interviews revealed the depth to which these classifications affected the students' reactions to the dissection process. One student from the first group explained, "To me the cadaver is a complete nonperson. You really don't think of it as being your body or somebody else's. It's just like a rubber model. When somebody says that the cadaver died of something, it sounds really strange. You don't think of it that way. I think it's pretty stupid to be squeamish with cadavers."

A quote from one member of the second group reveals a completely different mind-set.

One of the basic premises I have about lab is that you *should* have a reaction to it—the thoughts, the sights, the smells—because this is a dead person and you're going to be dead someday, too, and this is an inescapable association. There are people working in lab with me who never express their emotions. If they don't have that emotional sensitivity now, they'll be doing the same thing later on. There are going to be a lot of patients you are going to have to care for that will be physically, or whatever, unable to react to you, just like a cadaver, and you've got to be able to make yourself aware of the patient's feelings, his pain or discomfort, and acknowledge this as part of your work, something you must have if you're going to be a good doctor.

I discovered as the weeks of lab wore on that I am without question a card-carrying member of Hafferty's Group II. I can easily identify with the large proportion of students in the Stanford study who recounted stressful psychological effects. Reading the student vignettes included in the paper, three were particularly familiar to me and addressed the range of impacts that students faced. This description mirrored the ways in which Tripler was affected by lab:

The first student reported uncontrollable visual images that frightened her during routine daily activities. Anatomy laboratory was nearly intolerable, though her classmates seemed reluctant to discuss this with her. Patterns of sleeping and eating were disordered. Dreams included images of the inside of her heart, images of cadavers floating by her while she stood in chest-high water, and an image of her carrying a bloodied bone in a coffee can.

A second vignette reminded me of the disturbing and disruptive effects that dissection had on Lex:

Another student reported nightmares and poor sleep, including one dream where he pictured a "very sick woman in her late thirties, thin and limp and barely alive." She was pictured with cardiac hypertrophy and looked "mostly like a group of enlarged arteries pulsating in her neck, chest and arm." . . . He remarked that his "whole sleeping life had changed." After dissection of the face he had intermittent visual experiences for 6 months.

He repeatedly described "seeing through" the face of a pretty woman to her bones and muscles underneath.

A final description reflected the toll that anatomy class had levied on my own life.

This student experienced anxiety, insomnia, and depression during the first weeks of anatomy class. Concentration and study were difficult. She had images of "pieces of skin from the cadaver" while brushing her teeth.

These are the ways in which the stress of cutting into human flesh is made manifest: the tale of a toll-booth joke, the psychic echoes of trauma, dreams of cadavers floating by.

The written and practical exams are held on the same day, with the written portion administered in the morning in the lecture hall and then the practical in the lab in the afternoon, in two shifts. Some of my classmates hope for the second shift in order to have a few more hours of study time. I am happy, though, when, before the written exam, Dr. Goslow writes our table number as one of the tables in the first grouping. We will get it all over with as quickly as possible.

After a brief explanation of how the day will unfold, the written exams are handed to us. I scrawl my name across the top of each of the eight pages, take a deep breath, and begin to read.

Question one. *The drawing on the left depicts a cross section of a human blastodisc at the beginning of the third week; it is already a trilaminar embryo, but cells are still migrating. On the right, development has proceeded, and the gut tube is almost closed. On the right, draw in and label the following structures: neural tube, paraxial mesoderm, notochord, aorta, endoderm and intermediate mesoderm.* I draw in the structures quickly, label them, and move on. *One type of thoracic outlet syndrome is caused by compression of the lowest roots of the brachial plexus by a small, incompletely formed rib on the 7th cervical vertebra. Soft tissues associated with that cervical rib typically contribute to the compression as well. Indicate what nerve(s) will be affected by this compression and describe the symptoms that you would expect to see.* I write as fast as I can: "*C7 compression would impinge most severely on the posterior cord of the brachial plexus . . .*"

There are only a few questions at which I have to guess, and I finish the exam with enough time to look over my answers and add to the ones I finished in an early rush. When the exams are handed in and we're filing out of the lecture hall, reviews from my classmates are mixed. "That was miserable," I hear someone moan. "Yeah, I totally bombed it."

Trip leans over to me quietly. "Well?"

"I thought it was okay," I say noncommittally.

"Me, too!" she whispers, and then grins. "I know this is dorky, but I actually thought it was kind of fun!" I won't go that far, but I know what she means. Although no patients depended on our answers, we were solving clinical problems—and our ability to do so, for the first time, had consequences.

I discover, however, that the post-test feeling of well-being is short-lived. Moving on to the practical, my half of the class lines up in the hallway outside the lab, and as when we gathered here at the beginning of the semester, a quiet nervousness comes over us.

Dale swings open the door to the dissection room and steps out in the hallway. He explains that there are thirty-five numbered stations, with two rest stations interspersed. We will each walk to any random station to begin. The faculty will announce the start of the exam, and we will have ninety seconds to answer the two questions at that station. A beeping timer will sound, indicating that time is up, and we will advance to the next station. Dale hands out answer sheets with thirty-five numbered lines.

The dissection tables have been arranged end to end in a huge circle. There are cords strung up at various places in the lab and leading through the open door into and then back out of the pro-section room, to designate the one-way flow of traffic from one question station to the next. To prevent desiccation—and, surely, distraction—many of the cadavers have only small areas of their bodies uncovered. It dawns on me that I do not know whether I will recognize Eve, and that thought strikes me as very strange. Despite the reorganization of the tables, Trip and I automatically head for the far left corner of the room—the spot we occupy during each afternoon of lab.

Most of the stations have full cadavers on the tables, although there are some prosections that have been brought out. A single right arm lies on a tray. A few containers hold hearts in various stages of dissection. There are also a few stations with medical imagery—X-rays or CT scans or MRIs—and a couple of pictures. From across the room, I can see Michael Jordan's famous "Wings" poster, with his outstretched arms, and in the prosection room is an enlargement of one of the Vesalius woodcuts. Each image has colored arrows affixed to it for the exam, asking questions about the visible muscles indicated. The vast majority of the questions, however, involve our cadavers. There are two questions per station, and they are differentiated as red and blue. At the stations with cadavers or prosections, red and blue pieces of thread are tied around vessels or nerves, or red and blue pins are stuck into larger or more precise structures. Some questions ask us simply to identify what the marker indicates; others ask for more involved knowledge about the marked structures.

My first station is a cadaver, and says only *Red: Identify. Blue: Identify*. I look at the red pinhead, which is stuck in one of many muscles in the posterior section of the upper arm and back. With my gloved hand, I check its connections: the bicipital groove of the humerus and the lateral border of the scapula. It's teres major, which medially rotates and adducts the arm. *"Teres major,"* I write quickly, and look at the blue thread, which is wound around a vessel in a similar location. I gently touch the vessel and confirm that it's an artery: too sturdy for vein, yet with a perceptible hollowness, unlike nerve. I can't figure out whether I've got the main artery or a branch, and the timer sounds. "Move on, please," says Dale. I decide it's too small for the axillary artery and too low for the circumflex and, as I'm walking to Station 31, scribble. *"Profunda brachii artery."*

The questions at Station 31 are even worse. Red asks for the identification of one of the extensor muscles of the forearm, which seems to me to terminate in a totally different location from Eve's, a sure sign that it's I who am disoriented, rather than the musculature. A blue thread is tied around a nerve, and the blue question asks what intrinsic hand muscles are innervated by it. I know that it's the median nerve, and I know that all but one of the intrinsic muscles of the hand are innervated by either the median nerve or the deep ulnar nerve. What I can't remember is which grouping of muscles belongs to which nerve. The buzzer sounds, and with little confidence I write down, *"Palmar and dorsal interossei."* Hoping to have a chance to come back to the question at a rest station, I write above the answer, *"Intrinsic hand ms innervated by median n?"* In the end the answer I have written remains, and it is the wrong one.

When I reach Station 32, I'm unsure about three of the four answers that I've filled in. Therefore it's a strange relief to look down at the body on the table and see, unmistakably, that it's Eve. The red pin is stuck in her serratus anterior muscle and the blue thread tied around her spinal accessory nerve, a specimen that is so prominent on the back of her neck that Dr. Goslow called the class over to see it during lab one day. I write down the answers quickly and use the remainder of the time to take some deep breaths. I'll return to the intrinsic muscles of the hand at the upcoming rest station and hope that by then some buried knowledge will have surfaced.

The practical continues in this way. *What spinal nerve exits here? What is the function of this structure? Identify. Name the structure that occupies this groove. Identify. Name the last chamber of the heart that blood in this vessel was in. Does this structure carry somatic or autonomic fibers? Identify. Identify.* I know the answers to some questions immediately, others I try to sort out, and on some I draw a complete blank. All the momentum from the written exam has seeped away, and the only remaining push is from the adrenaline of anxiety. At the second rest station, the faculty has taped to an empty table a *Far Side* cartoon about a mad scientist with jars of brains in his secret laboratory. Trip has left me a tiny note on a torn corner of paper. *"Christine,"* she has written. I turn it over. *"Oh, dear,"* it reads, with a frown drawn beside it. I look up and find her a few stations ahead of me, ready to begin a question, and I smile. She shrugs and shakes her head as if to say, *Oh, well! Nothing to be done at this stage,* and then bends down toward the heart on the table in front of her, peering into it as if the answer that she seeks is written deep inside.

Leaving the exam, I pass Lex in the hallway on the way out. He's waiting to begin the practical's second round. "How is it?" he asks with a nervous cringe. "Over!" I say, and then, "You'll do fine." Despite the difficulty of the practical, I feel a great sense of relief and an instant surge of energy. I wave good-bye to Trip and drive straight to the market, where I buy a storm of beautiful and indulgent things: two fat and shiny butternut squashes from which to make soup, fresh basil and garlic to stuff into a handful of sea scallops, arugula and goat cheese; a bottle of Riesling, dark chocolate for a soufflé that always falls but is still divine, and for Deborah a bunch of cheery yellow, lovely-smelling freesia. With medical-school loans and one salary between the two of us, it is all a great splurge, but I spend the whole afternoon in the kitchen, chopping and sautéing and singing along very badly to Aretha Franklin. When I hear Deborah at the door, I run there with the dog and force her to drop her bag and stacks of papers in the entryway so that I can dance her past the vase of flowers and into the kitchen. I am delirious with relief and fatigue, but this night is free, and well earned, and I refuse to waste it.

When the exams are returned to us, I have received a nearly perfect score on the written portion and a narrowly passing grade for the practical. Tripler has scored slightly better on the practical, but neither of us is thrilled. To the whole class, Dr. Goslow says, "Scores on the practicals tend to improve over the term; it takes a while to get used to the system. But," he continues, "the practicals also tend to expose problems for folks who haven't spent much time with bodies other than their own. So make sure you're checking out structures on several different cadavers, because structures can really vary from one to the next, and it's a great way to be sure you know your stuff." Trip and I glance meaningfully at each other, having staunchly resisted—except our futile attempt to find the coronary arteries of the heart of Roxanne's table's cadaver—extending beyond our comfort zone from the neat and familiar territory of Eve. Dr. Goslow wraps up his speech by saying, "The other thing to consider, of course, is that the surest way to improve the practical score is to spend a little more time in the lab." A mass groan comes from the class, and Goslow grins. "I know, I know. Just something to keep in mind."

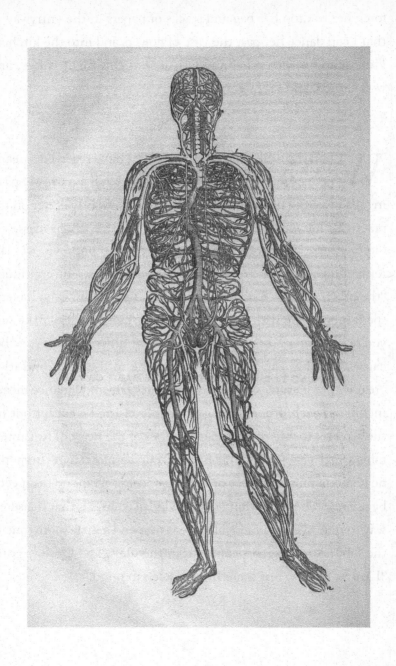

chapter nine

The Discomfort
of Doctoring

*What then did I formerly believe myself to be? Undoubtedly I
believed myself to be a man.*

But what is a man?

DESCARTES, *MEDITATIONS*

A nother dream: The body I am to dissect is underwater.
Her shape is not constant; the edges are rippled and
quivering from the creek flow's dappled light. The water gives her motion that does not belong in a body that is dead.
The creek is shallow, her body lying beneath the surface on water-
polished stones.

With my scalpel and forceps, I kneel beside her, the water rush-
ing around my waist, the current tugging me off balance. I make
tentative cuts, yet each opening I create becomes a new outlet

into which the creek can flow. Water courses through her femoral artery. Water eddies in her abdomen.

And just when I begin to think maybe this is beautiful, the current begins to carry parts of her away, splashing them over hidden rocks and pulling them along tree-lined turns. I am stricken with guilt. I futilely reach and reach in an attempt to contain the loss, but the water is strong and wide, and even what I manage to touch slips quickly through my fingers.

As her body empties, I feel more and more hollow. I think I must offer her some explanation, but when I look to her face, there is clear and perfect water swirling from her open mouth, a question in a language I cannot comprehend.

For reasons based in embryological development, the dissection of the male genitalia begins with partial dissection of the abdomen. In the sixth week of the semester, we are all instructed to make incisions down the cadaver's midline, from the skin remaining at the base of the thorax (veering around the umbilicus in every cadaver but belly button–less Eve; we can, of course, proceed in an unfettered straight line), and finishing by making a T at what my mother would call the "crotch." Although we know we will eventually dissect the female genitals, we are still reluctant to approach Eve's. We intentionally make our T prematurely, with the longitudinal incision stopping just before the body's pubic hair and the perpendicular incision cutting a line just above it, as if making it a long vowel. Just as we did with the thoracic skin, we open the abdominal skin like a book. This new opening makes the cadaver seem like the Dalí paintings of a woman who opens into drawers, or like religious paintings in which God opens his robe to reveal a multitude of stars.

Beneath the skin is a layer of fascia covering the abdominal muscles. The lab manual directs, "Note the superficial fascia (Camper's fascia) which contains variable amounts of fat."

"No shit," says Roxanne when she reads that line, gesturing toward our table. "They're already practically to the woman's backbone, and we've still got no clue whether or not our guy even *has* organs under all of this." To make matters worse, the cadaver at Roxanne's table has started to emit a strange, organic smell, exacerbated by the opening of his abdominal cavity. It will take two more days of lab, including the group's dissection and removal of all his internal organs, before Arnis is sure of what Roxanne claims to have known all along: The embalming process on their cadaver was incomplete, and it is rotting. "No shit," says Roxanne when they tell her. When they wheel the cadaver away and bring in a new, thinner one to replace him, she just stands, arms folded, and nods.

While we aren't quite to the backbone, as Roxanne suggested, we have cleared away the Camper's fascia and have uncovered the abdominal muscles. On a diagonal from your hip bones to your genitals, about an inch or two above the line of your underwear at the top of each leg, is the base of the abdominal musculature. At the base the edge of the musculature rolls under itself—like the rolled edge of an unfinished knitted sweater—to form the inguinal canal. This canal is present in both men and women but is particularly important in men, who are more prone to inguinal hernias.

These hernias can occur when contents of the abdomen dip into the inguinal canal. In the second year, the medical school will pay men to come to the hospital and lift their gowns for us as we awkwardly gather their scrotal skin with one of our index fingers and push the finger up into the scrotum and beyond it, into the inguinal ring, asking the patient to cough so that we can feel for the increased abdominal pressure that would cause a hernia to bump up against the finger. "I just ask that you don't make any sudden movements," my patient would say before I fumbled around and eventually found the canal with the man's assistance.

Because our cadaver is female, we pair up with another group in order to learn the anatomy of the male genitalia. A slit is cut in the outer side of each scrotal sac, and, in the words of the *Dissector*, we "free the testis and spermatic cord." We laugh a little at the language of liberation. Each testis can now clumsily flop around like an egg in a long length of pantyhose, and we all agree that they were probably better off nestled in their original tight quarters.

The penises, to a startling degree, resemble mushrooms. When I overhear one of my classmates confide to her tablemates that it is the first penis she has ever touched, I cannot help but think that this is a grim introduction.

Though the appearance of the male cadavers' genitalia is not especially impressive, the anatomy is actually quite remarkable. As evidence for the evolutionary favor granted to reproduction, the penis and testicles make a pretty convincing case. The external positioning of the testicles is a result of the fact that sperm are temperature-sensitive. High temperatures make for sluggish sperm at best, nonviable at worst; hence the sorority-house wisdom that in the absence of a condom a hot tub can be an (admittedly far less reliable) contraception alternative. The internal body temperature is too warm to be optimal for the health of sperm. The external, scrotal temperature is about three degrees Celsius cooler, and hence a testicular sac arguably conferred an evolutionary advantage over those with undescended testes.

In order to maintain the optimal temperature for the sperm, crazy anatomy has evolved in the male genitals. Because it comes directly from the heart, at the center of the body's core, arterial blood is a warmer temperature than venous blood, which is on its way back to the heart after having cooled down. This is the reason that our fingers and toes can feel cold long before the stomach and back do.

In order to leave the body via the penis, sperm must leave the testicular sac and enter the warmer realm of the body. Both the tubes in which they travel, the vas deferens and the urethra, run right alongside arteries pulsing with warm blood. The heat-sensitive sperm are protected, however, by a buffering coil of vein that wraps around each artery. In addition, the coil cools the arterial blood in order to prevent the blood filling the penis from warming the region too much.

Any discussion of the anatomy of this region requires unabashed forthrightness. This necessity sets in motion certain stages of acclimation in my classmates and in me. First we feign utter comfort. We adopt the attitude that liver and spleen and testicles are all similar body parts. Since the liver and spleen have no emotional resonance for us, the testicles shouldn't either. Not only does this immediately prove to be untrue, but it will remain so with less-predictable structures as well once we begin to encounter brain, tongue, and eye.

This first stage officially ends for me the minute I have to touch a cadaver's penis. Since our cadaver is female, I do not have to hold the penis to cut away its skin, separate the main muscle structures, and "free" the testes. Instead I can just inspect the already-dissected penises of other groups. But in a strange way I think I would have preferred to actually do the dissection. With Eve I feel an odd kind of intimacy. I try to remain tender with her, and respectful, even when the actions I perform seem to be a violation. She and my group members and I have a kind of shared history—we knew her when she was whole. And so when we walk up to the man on the table next to us while his dissection group is at lunch, unzip the bag that covers him and reach immediately for his penis without regard to the flexors of his fingers or the arch of his aorta, without knowing the contours and quirks of his body at all, it feels like a violation. It feels like what it is—touching the penis of a stranger.

The second stage of acclimation is a kind of paralysis. This is *not* comfortable, no matter what I pretend, so I should do it as little as possible. I give the cadaver's penis a quick lift. "Okay," I say, looking at my classmates' careful dissection, "this must be the corpus cavernosum, and the corpus spongiosum here, and there's the spermatic cord"—I wave a gloved finger in its general vicinity—"and there are the testes. Do you see the epididymis, Trip?" I ask, sounding confused but knowing that it is found on the back of the testis and that it will surface as soon as she rolls it over. I am just unwilling to drop the shaft of the penis to reach over and give the testis a spin. I am desperately trying to avoid appearing freaked out, while simultaneously trying to avoid touching any more of this man's genitals than I have to. We all do our own versions of this balancing act, and after about thirty seconds, having gained little to no knowledge, we zip the bag back up and head back to the familiar shape of Eve.

The discomfort I feel is not without historical precedent. In an exhibit of antique medical instruments at Brown University's John Hay Library, viewers' eyes are naturally drawn to a lovely painted, ivory figurine of a naked Asian woman. She lies on her side, one arm above her head, the other bent at the elbow and holding a flower. Her hair is upswept. She is bare but for the shoes covering her disproportionately tiny, once-bound feet—a body part considered too private to be revealed even when all else is. She is a diagnostic doll, relied upon by seventeenth-century Chinese doctors in an era when it was deemed improper for a lady's body to be viewed, let alone examined. And so a lady patient would point to the area on the doll's exposed body that corresponded with the pain or problem she was having, and the doctor would make a diagnosis on that basis. Viewers surrounding the exhibit gasp at this and laugh, and they should—pointing to the belly could mean anything from a ruptured appendix to pregnancy to constipation. But when I am charged with studying the penis of a cadaver, I understand how diverting attention to an unreal substitute does have its allure.

The final stage of acclimation to violating the norms of bodily privacy takes some intellectual work. And, in all honesty, I think as a result it is a stage not universally reached. At the very least, it comes in gradients over time. When I go home on the day of the penis dissection, I feel ashamed of myself. It is not some kind of puritanical shame related to genitals; it is the shame of inadequacy. I know that, as patients, our levels of comfort are often direct reflections of the levels of comfort our physicians exude. A gynecologist who addresses issues without embarrassment or evasion instantly puts me at ease. By contrast, I once went to a cardiologist who blushed each time I opened my shirt for him to listen to my heart. His response provoked me to feel uncomfortable and ashamed, and as a result I neither liked nor trusted him.

The afternoon of the penis dissection, I come to understand that I *will* have to touch the penises of strangers soon, and the breasts, and the wounds, and the fat, and the growths, and not those of dead people. They will be those of patients with keen perceptive ability, ready to pick up from me any sign of disgust or discomfort. It should not be the responsibility of sick patients to bear the burden of unease, I realize, and if I won't be able to exude comfort, I won't have much of a chance of being a trusted and well-liked physician.

In my first clinical month in the hospitals in my third year of medical school, I will have many opportunities to practice becoming at ease, with varying degrees of success. Many male patients express a kind of apologetic sheepishness on physical examination or history taking from a female student. "I feel funny saying this in front of you," a handsome, athletic twenty-six-year-old says to me. He had told the nurse that he had come to the doctor because of a headache, but, he confesses, "I really came for this rash on my ass."

For penis and testicular exams, I decide to always ask a male colleague to be present in the room, which makes the patient either less apt to admit his embarrassment or less prone to feeling it. However, by myself in the outpatient office or emergency room, I often have to ask a patient to unzip his pants and pull his underwear partially down to fully expose his abdomen. At this, several men, particularly of middle age and older, balk. "Really?" they ask.

On just such an occasion, one elderly man says to me, "You're going to be a woman doctor, right?" Bewildered, I answer yes, wondering what the alternative could possibly be until I realize that he is gently suggesting that I become a gynecologist.

There is a documented phenomenon clinicians call "white-coat hypertension"—that is, blood pressure that rises in the doctor's office purely due to nervousness. The female residents with whom I work joke about a kind of rapid heartbeat called "young-female-doctor tachycardia."

I discover that whenever a patient expresses any sentiments of embarrassment, I feel comfortable saying, with some slight variation, "Oh, gosh, Mr. So-and-so, I examine all kinds of people every day, but I'm happy to step out and send in a different doctor if that would make you feel more comfortable." No one ever takes me up on it.

For a doctor-in-training, learning to be at ease with a patient's body is not the same as learning to be at ease with a patient's fears about his body. The midpoint in medicine between excessive emotional involvement with patients and a complete lack of empathy is not a simple one to locate. The former leads to exhaustion and burnout in the care provider; the latter gives rise to the all-too-familiar doctor with "bad bedside manner," whose patients feel unheard, uncared about, and, as a result, unsafe. The attempt to define a balance between the two has led to that ideal of detached concern, in which clinicians should demonstrate concern for their patients but should detach themselves from any emotional response that their patients' condition might evoke. The premise is that a physician's clinical judgment might be compromised if he were to become emotionally involved. I would learn, however, that there is also a component of detachment that benefits the patients and their families in less directly clinical ways.

I was in my initial two weeks as a third-year medical student on the hospital wards when I had a reaction to a patient which was charitably described as ineffectual by the senior resident teaching me, but more accurately characterized as inappropriate. The very first person who was ever assigned to me for (carefully supervised) care was a fifty-two-year-old man with lung cancer that had metastasized to the brain. He came to the hospital with increasing leg weakness. The patient, Mr. R, had recently completed a massive series of radiation treatments to the head, and the scans of the brain had shown no tumors. He had only just received that news, and he was ecstatic. In the emergency room with his sister and his teenage daughter, he was eager for us to send him home.

In fact Mr. R was admitted to the hospital, where he stayed under my care for nine days. One by one, the senior physicians, residents, and I ruled out potential causes of the leg weakness. His spinal cord might have been compressed by a new tumor, but the MRI showed no such growth. Perhaps the steroids he was taking to slow the cancer were causing the change, but tapering down the dosage of prednisone had no effect. We consulted neurologists and oncologists to no avail, and in the meantime Mr. R stopped being able to move his bowels. He was suddenly afflicted by excruciating pain in his abdomen and a spike in the number of his white blood cells, indicating an infection. A new scan of his belly showed uncharacteristic brightness in part of Mr. R's small intestine, suggesting that the blood to a portion of his bowels had been blocked. His intestines in that area might be infected, or even rotting. A surgeon was consulted, but by now Mr. R's pulse and blood pressure were out of the range considered safe for surgery. In addition, he and his family, so recently celebrating the news of his cleared brain metastases, were now questioning how much sense major surgery made in light of these new problems.

By the eighth hospital day, Mr. R's abdominal pain was so bad that he had to be heavily medicated; he had moments of lucidity but was terrifically fatigued and heavily sedated. We had been unable to determine the cause of Mr. R's new crisis, and it was growing increasingly unlikely that we would ever do so. At this point the oncologist joined me to speak to Mr. R and his family about their diminishing options. The first was to keep Mr. R alive at all costs, even if it meant his being hooked up to a ventilator or defibrillated. The alternative was a less-aggressive solution that would essentially involve measures of comfort, easing Mr. R's transition toward death.

The oncologist had encouraged me to take an active role in the conversation, and my senior resident, who had cared for Mr. R on many occasions over the course of his illness, agreed. He said that the purpose of my clinical training was in part to become comfortable in circumstances like these. I knew all three of Mr.

R's sisters by name and knew his three daughters as well; I also had studied the case closely over the last eight days and so felt as if the discussion were a natural one for me to begin. I explained to the family and to Mr. R, who was drifting in and out of the conversation, that we perceived Mr. R's body to be failing. I tried as best I could to explain the options and the decisions they had to make. The resident and oncologist gently entered into the conversation at times to clarify questions the family members had, and then they stepped back, indicating that I should continue. When I reached the end of the explanation, there was a long stretch of quiet, with all six women crying.

Finally the eldest daughter, twenty-two, to whom her father had assigned his power of attorney, lifted her head and through her tears earnestly asked me, "If this was your father, what would you do?"

I began to cry. Not an audible or sobbing cry, but a cry that made my voice quiver as I tried to respond. "It's difficult . . ." I began to stutter. Before I could continue to grasp for words, the oncologist stepped right in front of me and began to speak calmly of the reality of the situation, of the strong and real feelings that would naturally and rightly accompany any such decision. He reviewed each option, gently fielded questions, and exuded a quiet compassion. All the while he steered the family gently toward a decision.

As I stood in the recesses of Mr. R's room watching the oncologist speak with simultaneous warmth and professional distance, I realized that when the daughter asked me what I would do, what she needed most was for me to be clear. She needed clarity in the midst of her emotional turmoil.

I could not have known, as I stood over the male cadaver being dissected by our neighboring group, that the awkward moment of looking at a dead stranger's genitals would be comparable in some ways to discussing end-of-life issues with a patient's family in the future. I did know, however, that any discomfort I was feeling in the anatomy lab was interfering with my ability to learn important medical information. And whether my future patients would require clarity or knowledge, my ability to manage my own discomfort in the face of their bodies and their illnesses would be one of the most critical lessons of my medical training.

On the far opposite end of the spectrum of clinical detachment are those who distance themselves to an extreme degree. Historically, it is this severe detachment that must have led to the mistreatment of cadavers documented over the centuries. From the resurrectionist's cart to the dissecting table, bodies were not treated gently. In the 1800s one Dublin professor published scientific observations about physical changes in the brain resulting from fever. Because his studies had been conducted on cadavers, his findings were strongly contested. The basis of the challenge was that corpses were handled so roughly during disinterment and transportation, it was impossible to determine which of the changes in a cadaver's brain were attributable to fever and which were a result of trauma inflicted upon the body after death. The objection stated, in part:

The bodies, tied up in a sack, neck and heels, are subject to great violence during their transport from the Burying grounds, for example, they are dropt from a high wall, by which it is well known fractures of the cervical vertebra not unfrequently occur. . . . The cadaver is roughly handled in the dissection room, being often left with the head hanging over the table, by which means the blood gravitates, and congestion of the brain is the consequence.

Indeed, this kind of violent treatment in transporting the corpse was not at all uncommon. As Ruth Richardson chronicles in her fascinating *Death, Dissection and the Destitute,* the bodies of a deceased woman and her baby were once found crammed into a box two feet by two feet square, with each side nailed shut. Richardson continues:

Human bodies were compressed into boxes, packed in sawdust, packed in hay, trussed up in sacks, roped up like hams, sewn in canvas, packed in cases, casks, barrels, crates and hampers; salted, pickled or injected with preservative. They were carried in carts and [wagons], in barrows and steam-boats; manhandled, damaged in transit, and hidden under loads of vegetables. They were stored in cellars and on quays. Human bodies were dismembered and sold in pieces, or measured and sold by the inch.

Hogarth's famous engraving *The Reward of Cruelty,* on display in University Museum, Oxford, shows a dog beneath the dissecting table, ravenously biting into the corpse's heart, and Jacques Gheyn's engraving *The Anatomy Lesson of Doctor Pieter Paaw,* features two dogs in the forefront of a similar anatomy theater scene, like vultures portentously awaiting their due turn at the body. These representations are unlikely to have emerged from the artists' imaginations, as "keepers of wild beasts" were said to have eliminated the need for anatomists to bother themselves with the reburial or disposal of cadavers once their dissections were finished.

Some of the most incredible and galling accounts of the treatment of cadavers, however, involve the numerous implied, and occasionally explicit, descriptions of the sexualization of—particularly young female—dead bodies. One antianatomy speech in the nineteenth century alluded to a sexualized impropriety in the labs:

> Who . . . even among the practitioners of medicine, does not shudder at the mere contemplation that the remains of all which was dear to him, of a beloved parent, wife, sister, or daughter, may be exposed to the rude gaze and perhaps to the indecent jests of unfeeling men, and afterwards be mutilated and dismembered in the presence of hundreds of spectators.

Ruth Richardson describes a cartoon to the same effect, which shows a woman in a London anatomy school on Resurrection Day. The woman demands "the return of her virginity, apparently lost since her arrival there."

Though I hope that even the greatest sense of personal distance from a cadaver would not have led me to act in the slightest way that compromised a body's integrity, does my response to Eve—compared to how I might have felt dissecting her executed or impoverished double—bring to light a potential shortcoming in me as a health-care provider? As was the case with the daughter of Mr. R, do I need to be able to see myself in my patients in order to treat them with the appropriate degree of empathy and esteem? Partially, I am beginning to learn that the answer is yes.

In my second month on the hospital wards as a third-year medical student, I am assigned the patient in Room 412. When I round the corner of the hallway and raise my eyes from list of patients I'm carrying, I see that a police officer sits in a chair in the doorway of 412. I greet him, slip past the chair, and introduce myself to my new patient, who is a thirty-eight-year-old man with a huge tumor in his neck. He has been transferred to the hospital from the state penitentiary, and one wrist and one ankle are handcuffed to the metal bed rails.

I take a social and medical history, which for this patient includes many stints in prison, early departure from school, a spotty employment record, heavy drug use, and positive HIV status. He and I could not be more different, and yet as I perform his physical exam, he jokes easily about how much he prefers even the oft-maligned hospital meals to his daily prison fare and how, tumor or no tumor, the nurses are nicer to be around than the wardens. I am not so naïve to think that his persona is as simple as all that, but despite the difference in our positions, it is not difficult for me to see him as he is: a sick man who needs me to provide him with the same treatment I would give any patient, with or without handcuffs, with or without HIV. I cannot legally ask him why he is in prison, but because of the increased risk of inmates having hepatitis or HIV, of using IV drugs and having unprotected anal intercourse, it is relevant to his medical history how long he has been in prison. Therefore, when he tells me nineteen years, I cannot know the specifics of his crime, but I do know he may have done something truly awful. Yet unlike the dead body of an executed murderer, my patient's humanity persists in a way that makes it impossible for me to mistreat him as early medical students may have done to their criminal cadavers.

I am aware that this, too, is a nebulous boundary. I may congratulate myself on the fact that I can care for my prisoner-patient without hesitation, just as I congratulate myself for regarding Eve as something more than a biological specimen. Yet my empathy is not always so readily available. Two years after anatomy, during a rotation in the surgical intensive-care unit, I am responsible for patients who are technically alive but who are, due to sedation or grave illness, partially or completely unresponsive. Some of them are the wrong color, some are terribly bloated due to fluid retention and, even in the absence of any disfigurement, no longer resemble any person you or I have ever seen. I sail into one of these rooms, call out "Good morning, Mr. C!" with false cheerfulness, and say loudly, "It's Christine, the medical student. I'm just here to do a quick exam, okay?" Mr. C, of course, does not answer. Could not answer even if he were otherwise able, given the tubes in his throat and nose, the ventilator beside him delivering measured breaths to his lungs.

I narrate my exam into the room's silence: "I'm just going to take a quick listen to your heart and lungs. I'm just going to press on your belly." I lift his gown and only glance quickly at his soft-ball-size testicles—a shared trait of many of the sickest and most edematous men in the ICU—then say, "Just going to change your bandages!" I hear that all my sentences have "just" in them, as if

I am trying to come and go unnoticed, as if the position Mr. C is in is no different from that of a schoolkid on picture day. *Just sit still and smile. This will only take a second.* I hold my breath, and peel away tape that pulls at hair and skin to free the patches of gauze on Mr. C's abdomen. Some of the bandages are simple squares that wrap the rubber tubes extruding from his body, draining pus, or blood, or fluid, or sending liquid food into his stomach. The tube sites ooze, and the gauze absorbs it, and I change the gauze, wipe him clean, secure new squares in place with tape. The center incision from some procedure that preceded me gapes broadly, infected, and refuses to heal. It is a foot long, leaving Mr. C's belly looking like a huge, split loaf. The split is beefy red and packed with gauze. A thick, greenish liquid soaks the gauze and bubbles out of the incision site when I press on Mr. C's stomach. It has a horrible, acrid, rotting smell that makes me cough and gag. I breathe out of my mouth, mop the wound with more gauze, throw away the saturated wad, and repack the wound with reams of fresh gauze, a new bandage, more tape. I exhale, pull the johnny back over the poor expanse of his body, and say "All done, Mr. C. I'll see you tomorrow!" And through the whole thing, I feel nothing for Mr. C. I feel a little nauseated, desperately in need of fresh air, and a little sorry for myself.

This goes on for three days, and I dread the chore of changing Mr. C's bandages. I enter his room, give falsely cheerful greetings, a cursory exam, and as quick a change of gauze as I can manage. On the fourth day, everything changes. Mr. C is in exactly the same condition; he is no better, no more coherent. His wounds are just as raw and infected. But when I enter, I see that markered drawings have been taped to the wall of his room. A photograph of a man, woman, and two elementary-school-age boys has been taped to the bed rail, in a place where, if Mr. C would only open his eyes and turn his head, it would be directly in his field of vision. No matter how closely I look, I cannot believe that the grinning young father in the picture is Mr. C, the same man who looks so little like a man, so much less *this* man, and yet I know it to be true. On his headboard, in a childish, crayon scrawl, is a paper sign that, intentionally or not, is unquestionably for me. In capital letters, it reads: MY DADDY IS GOING TO MAKE IT. HE PROMISED ME HE WOULD.

Like the cadaver, Mr. C is voiceless and cannot assert his humanity to evoke empathy. Unlike Eve, in this sick and mechanized state, he resembles nothing I have ever loved, even distantly. The lesson that he teaches me is that I am not so unlike those medical students who maligned their criminal corpses and disrespected their pauper bodies. Am I, perhaps, worse? The cadavers, after all, are in essence beyond harm; the only injury they can suffer is to their dignity, in whatever form it still exists. Conversely, despite Mr. C's imprecise state of existence, it is clear that the injuries my lack of care could cause him are real. The lesson he teaches me is that medicine's charge demands that we must not *need* voices or jokes or photographs or physical resemblances or shared backgrounds or heartbreaking crayon signs in order to treat our patients (and their bodies, when, in life or in death, the humanness of them seems elsewhere or gone) with tenderness and empathy and honor.

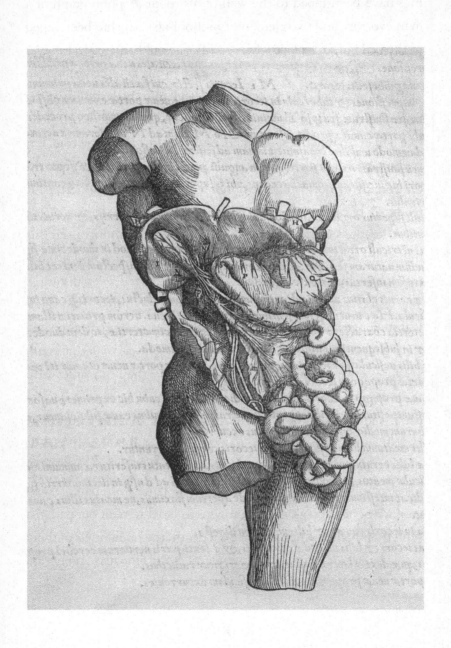

An Unsteady Balance

You are a little soul carrying around a corpse.

EPICTETUS

Although we've finished the dissection of the male genitals, we will not study their female counterparts for two weeks, so that they can correspond with the uterus in the pelvic dissection. In the interim, on the second day of the abdominal dissection, we open Eve's abdomen more fully to examine her internal organs. The mood in the lab room is light, with the stress of the first exam behind us and the transition from male genitals to guts a welcome step down in emotional intensity. Although the initial cuts into the abdomen have removed the skin and exposed the abdominal musculature, we still need to cut through the abdominal wall. Trip grudgingly volunteers to be the one to make the incision, and once the belly is opened, all we see is a chaotic mess of fat and intestines. Raj is reading our instructions aloud from the *Dissector* and at first doesn't raise his head from the pages to see what we see. He reads the names of the discrete organs and anatomic landmarks we're supposed to find: "Okay, identify the fundus, body, antrum, and pylorus of the stomach. Then find the hepatogastric and hepatoduodenal ligaments of the lesser omentum, noting the greater omentum with its gastrocolic and gastro - splenic portions." Tripler begins laughing so hard that she is practically in tears.

"Oh, yeah, I see the distinction between the gastrocolic and gastrosplenic portions quite clearly here, don't you, Christine?" She gestures to the indistinguishable innards and can barely get the whole sentence out. She has to force herself to stop laughing in order to breathe.

"And here is the greater curvature of guts," I respond, giggling, "followed by the lesser curvature of chitlins and the ligament of tripe." For a moment Raj is disoriented, searching frantically for the names of our newfound structures in the *Dissector*. When he looks up, he breathes a deep sigh of frustration while waiting for Trip and me to compose ourselves and resume.

The room is full of discussion about how jam-packed the abdomen is with viscera. Two students from a table across the lab each take an end of the intestines they've cut free from their cadaver and step back from each other until the intestines are fully extended between them, like a clothesline. They are at least ten feet apart. We all stare at them, half amazed by just *how much* intestine there is and half stunned by the shift in all of us that has made it permissible—even natural—to do something like string out someone's innards to see how long they can stretch. A third member of the table bends his body backward and walks toward the line of intestine, as if he were about to limbo beneath it. He stops and laughs with no intention of going through with it, a perfect expression of the tenuous balance of appropriateness we are beginning to find—of the transformation we are in the midst of negotiating: *I will nod in the direction of humor, of absurdity, but I will stop myself before I cross the line into disrespect.*

We do eventually find Eve's stomach, which, paradoxically, is the largest of any of the cadavers' in the lab. Like the others it is a flat sac, pale pink and nondescript but for the thick, ringed pylorus, from the Greek word meaning "gatekeeper." The pylorus tightens and relaxes, determining when the gastric contents can pass into the intestines. Though the interrelationship among the stomach, pylorus, and intestines is straightforward enough, the abdomen introduces a host of complex anatomical intertwinings. The green fist of the gallbladder stores bile made in the liver and releases it into the duodenal portion of the intestine via a pathway of tortuous ducts. The broad portal vein brings blood through the liver for filtration with its multiple anastomoses. A seemingly endless number of arterial branches provide the gut its blood supply. The network of structures and names is utterly alien to us—we will not learn the functions of many of these organs for weeks yet. We have no systematic understanding either, of the ducts and veins and arteries and lymph nodes running through the region.

Confusing as the complexity of the abdomen is, we soon learn that the story of how it forms is even more mind-boggling. Dale delivers the lecture on the fetal development of the abdomen. He hands out a series of color diagrams that are sequential pictures of stages of what he has humorously—but also pretty accurately—called "The Dance of the Intestines." In it we learn that in order for our esophagus, stomach, intestines, liver, and pancreas to form and position themselves correctly, buds of organs come off the foregut, a precursor to the digestive tract. These buds "migrate," literally sliding from one area of the body to another, and, in the case of the pancreas, fuse with buds elsewhere to produce the juvenile form of the developing organ. The stomach then takes shape and begins to rotate. It also begins to grow a first intestinal loop. This loop of intestine actually *leaves* the fetal body, rotating 90 degrees as it pushes out into the umbilicus. There it continues to grow and then undergoes an additional 180-degree rotation as it returns into the fetal abdomen.

It makes no sense. Dale goes over it multiple times in lecture. He points to the picture representing each stage as he explains the directions of the rotations, the exit, the return. Lex and I look at each other, baffled. When we consult our embryology textbook for an in-depth explanation, we are met with horrific pictures of fetal abnormalities that occur when this rotation goes awry. Our friend George, a frighteningly smart guitar player, suggests we meet at his house that night with Play-Doh, pipe cleaners, and beer to try to reconstruct roughly what happens. We agree, figuring that seeing the evolution in three dimensions and making the movements with our hands may help us understand it.

In the end we spend a few hours accumulating our supplies and trying to mold organs out of Play-Doh, only to make no more sense of the embryology than when we began. We decide instead to cook omelettes and drink Guinness while going over our notes from the week. We make a trivia game out of it. I name a structure in the abdomen, then ask Lex a question about its embryology, or innervation, or blood supply. Lex answers—if he's right, he asks George another question about the same structure; if he's wrong, George tries to answer the original question correctly, then asks me a different one. We go around and around with as many structures as we can think of. Lex knows the answers to twice as many questions as I do, and George knows twice as many as both of us combined, but the atmosphere is silly and light. Before Lex and I leave, we try to go over Dale's pictures one last time, to no avail, and then head home.

Dale's drawings force us to begin to think about the new ways in which we are called upon to visualize the body. Obviously, when we are with our cadavers, we look at them mostly from only one perspective. We look down into their newly opened spaces and study what we see. We may pick up the stomach or the heart and turn it in our hands, but we must learn to see the insides of the body from multiple perspectives, an ability that our dissections cannot help us acquire.

In another perspective, the cross section, the body is drawn or viewed in many slices, like a loaf of bread. Learning cross-sectional anatomy is an important skill, which we will be called upon to use in our clinical rotations. We will need to look at CT scans of brains to try to determine, on the basis of those cross-sectional views, whether the ventricles are enlarged or whether there is blood beneath some of the brain's protective layers. We will need to make decisions about whether a patient requires surgery based on cross-sectional images that show inflammation, tumors, or some other abnormal anatomy. But the perspective we get on the body from dissection alone does not allow us to envision the relative position of one vessel to another—even one organ to another—in cross section.

In order to teach us to read cross-sectional views, Plexiglas squares are scattered on lab benches all around the periphery of the room. Cross sections of an actual body are embedded in these squares. They are stored in a back room, where they are kept in museum drawers that might elsewhere hold ancient maps or pages from illuminated manuscripts. Dale and Dr. Goslow bring out several squares at a time that correspond with the part of the body we're studying. Rounds of bone are surrounded by rounds of muscle, like grayed cuts of meat. The sections offer a helpful view, bizarre as it is, of how all of the structures interconnect, no organ or muscle working in isolation. In the midst of the abdominal dissection, which can feel like a disorienting mess, the squares help bring order to the chaos.

The strangest Plexiglas sections are those of the head, because here the body's recognizable exterior features refuse to hide. The sections are numbered, beginning with the uppermost cut of the scalp as section one. Section number four reveals a beautiful view of the brain's tissue, vasculature, and ventricles but also contains the eyebrows. Section five shows a look into the sockets of the eyes and the beginnings of the cerebellum, but most striking are the halved ears that jut out on two sections. In lab one day, I pick up each of the head sections, and look at them not from the top, as we are meant to, but from the side, where the facial features are, in inch-by-inch increments. Here the eyelids and lashes, here a section of a nose, pores visible, hair in the nostrils. Here the lip in front of the slice of jawbone, of flat, pink tongue. Here whiskers.

I wonder about this man—we first understood he was a man from the relative absence of breast tissue in his thoracic sections and later from the sliced segments of his penis and testicles. Some of my classmates had a guess even earlier, when they picked up a Plexiglas piece and looked at it from the side, seeing many dark hairs on the preserved surface of the skin. Could he have possibly *chosen* for this to be the way in which his body was to be used? And why does the way his body has been sliced and preserved strike me as more disconcerting than the cuts and the parceling we are doing to Eve? Is it simply that I have become accustomed to the process of dissecting the body as a whole, I wonder, a whole that retains— at least up until now—its human shape? Or is there something reassuring about the knowledge that Eve will eventually be cremated, will be delivered from this state in which I am keeping her? Is it the permanence of this sectioned man that so disarms?

I ask Dr. Goslow where these Plexiglas sections come from; how does one go about purchasing a man in one-inch increments? As is turns out, the answer is that you send a cadaver and twenty-two thousand dollars to a guy named Davy in Kentucky, who will embed the body in plastic and then cut it into slices with a band saw.

As the abdominal dissection progresses, we discover that Eve does not have a gallbladder. Unlike the absence of her umbilicus, this does not raise our eyebrows. Instead of assuming some mythical origin as an explanation, we assume that she has had her gallbladder removed, a cholecystectomy, a procedure we will see many, many times in our surgical rotations. Patients will come in with sharp, severe pain along their right sides, and we will accompany them into the operating room to "drive" the laparoscopic camera the surgeons will poke into their bellies, then watch the green sac be clamped off from its blood supply and bile ducts and removed from the body for good. Sometimes the sac will break and the thick, green-brown ooze of bile will spill into the patient's abdomen, perhaps with small white stones floating along, and the surgeons will have to thoroughly rinse the abdominal cavity (an action they poetically call irrigation, as if they were encouraging crops to grow) so that the bile doesn't irritate the other abdominal organs or the walls of the cavity.

The eerie, alien appearance of bile allows one easily to understand its historical demonization as a bodily fluid with the capacity to wreak havoc on the balance of health. In fact, bile was known as "choler" until the eighteenth century, from the Greek *cholera*, which the *OED* defines as "the name of a disease, including 'bilious' (jaundice) disorders." The Greek *bile* meant "bitter anger" and became the ordinary name of one of the four humors (cholera, melancholia, phlegma, sanguis) whose tenuous balance was said to dictate the state of one's health and spirit. Initial theories held that in Adam and Eve before the Fall the four humors existed in a precise balance. Mortality, however, brought with it a perpetual imbalance, in which one humor was always in recognizable surplus and another therefore in deficit. Bile was said to be the one of the four humors that caused "irascibility of tem-per." Chaucer is responsible for the first documented use of "choler" in 1386:

> Dreams are engendered in the too-replete
> From vapours in the belly, which compete
> With others, too abundant, swollen tight.
> No doubt the redness in your dream to-night
> Comes from the superfluity and force
> Of the red choler in your blood. Of course. . . .
> For melancholy choler; let me urge
> You free yourself from vapours with a purge.
> . . . And purge you well beneath and well above.
> Now don't forget it, dear, for God's own love!
> Your face is choleric and shows distension;
> Be careful lest the sun in his ascension
> Should catch you full of humors, hot and many.
> And if he does, my dear, I'll lay a penny
> It means a bout of fever or a breath
> Of tertian ague. You may catch your death.

What about this tension of dreams, to which Chaucer alludes? This need to purge oneself lest the imbalance of vapors brings unto our own selves literal or figurative illness or death? Is this the role that humor plays for us—stretching intestines across the room? How does one absorb those things we are not meant to ab-sorb, and maintain an evenness? How to swallow the "melancholy choler" of other people's sickness and death, to provide strength and support and knowledge to our patients, and yet detach appropriately so that we can lend ourselves fully to the matters of our own lives?

When we finish the abdominal dissection, we can finally study and name the contents of Eve's abdomen; we understand bile more fully than Chaucer did. And yet some aspects of doctoring—the turns of the fetal gut outside the body, a man preserved in Plexiglas slices—still feel as vaporous as humors.

In the semester that follows anatomy, I will request to observe an autopsy in my pathology course. Having spent months in a room full of cadavers and having taken the body from a whole to its parts, I have no apprehension about attending the autopsy. However, when I enter the morgue, I find I can hardly stand. The body under examination is a young woman who is presumed to have died from complications of multiple sclerosis. Unlike our anatomy cadavers, she is young, has a full head of curly auburn hair, and wears a wedding ring. As you and I would be, she is also still full of blood, and as the autopsy technician makes a sweeping, Y-shaped cut across her chest and abdomen, a hose runs water beneath her and onto the table to course away her blood.

In the cavity of her body are the same organs as Eve had, but their unembalmed condition means that they are more slippery and more flexible than Eve's stiffened liver, stomach, and intestines. The surfaces of every organ are covered in a deep red sheen of blood.

The most difficult aspect to tolerate is not looking at this woman's body but smelling it. She had been found in her home as many as twenty-four hours after she had died, and at the time of autopsy her body's natural decomposition has already begun. The stench is overpowering, and as her body is opened and organs are removed, the potency is multiplied.

As I stand reeling, while hoping to appear as if I were watching attentively, I will my legs to support my weight. The technician and the pathologists go busily about their work. They weigh and section each organ, looking for visible abnormalities, slip tissue samples into vials for microscopic examination, call out measurements to be recorded, saw open the skull and spine, remove the brain and snip spinal-cord samples, pack the body with sheets to compensate for the vacancy left by the removed organs, replace the skull, and finally sew the trunk and scalp back together with wide stitches. The skin stretches taut with each pull, like hide. The stitches on the head are hidden by hair. If it was not for the Y seam on her naked form, you would never know that this body had been disturbed.

Through all of this, the doctors ask kind, everyday questions of the technician and of each other: *How is your daughter's baseball team faring? When is the continuing-education lecture today?* They joke about the backlog of autopsies to be done and the rate at which Rhode Islanders seemed to be dying; they also joke about who should stay behind after the autopsy is finished to flush the blood and specks of organ and flesh into the wide, stainless-steel sink.

One gets accustomed to one's job, and the normal range of human emotions needles its way into any working space. For my classmates and me, we have begun to realize that this desensitization is a vital part of our rite of passage. We must learn not to be frozen in place by the sight of death, by the sight of nakedness, by disease and the body cut open. We will be called upon to act calmly and ably when we are confronted with the body in crisis. We must begin to individually shape how we will cope with the impact of these extraordinary moments of urgency and whether those coping mechanisms take a healthy or unhealthy form.

I will more fully appreciate the incremental fashion in which we are introduced to the body in trauma two years later when, during an elective pathology course at the Royal College of Surgeons in Ireland, I participate in autopsies with the histopathology staff. My first is a twenty-three-year old kid who had died the night before in a motorcycle accident. Though I witness several autopsies over the course of my time at the college, this is the one that distresses me the most. Willy, the autopsy technician, tells me how the young man died: A metal post had been bent down so that it extended into the road, and when he rode by on his motorbike, the post hit him in his side. When the medics arrived, he was standing, telling the medics what happened. He had a few scrapes and said that his side and shoulder hurt, but he wasn't bleeding and had no visible serious injuries. The ambulance drove him to the hospital, and when it arrived at the emergency room, the young man was dead.

I'm working with the Nigerian pathologist, Ade. He takes me down to the postmortem suite, where he gives me scrubs to change into and a pair of once-white, bloodstained rubber boots. He also hands me a surgical bonnet, gown, gloves, and a green plastic apron. Once properly suited, we enter the autopsy room, and I'm aware that I'm subconsciously holding my hands in front of my chest, being careful not to brush up against anything in the room, as if we were about to perform surgery and needed to avoid contamination.

Ade laughs. "This isn't the OR, you know." And I do, but these encounters with the newly dead can be jarring enough to displace reason. Once, during a different autopsy, I leave the postmortem suite to look for a ruler—the elderly dead man had an unruptured aortic aneurysm, and Ade needs to note its diameter. When I return and swing open the door, I see the old man's face and I am startled. I had forgotten that anyone but Ade was in the room.

The autopsy of the young man is well under way when we enter the room, and Willy is finishing up. At first I breathe through my mouth, trying not to fully take in the smell. The downside of this tactic is that you never grow accustomed to the odor—the sensory receptors of olfaction never fill. So, after a while, I opt to gradually begin breathing in through my nose, and before long I notice the smell only when new cuts are made or organs are moved, the slick red-brown glazing the stainless steel.

The young man's thorax and abdomen have already been opened by a long cut and the bone saw. Willy grasps and loosens the trachea and esophagus at the base of the throat, making a clean cut through both tubes. He grips the portion below the cut and pulls it firmly in the direction of the body's toes. The trachea and esophagus pull up and out, and the structures attached to them follow: the stems of the bronchus and the floppy lungs into which they lead, the heart with its pericardial sac and its great vessels, the stomach, the pancreas, the spleen. Willy uses his scalpel to cut away any connective tissue holding fast to the body and then cuts the organs, removed en bloc away from the intestines, which he also cuts at the level of the rectum and removes. The intestines go into a basin in the sink.

The organs are set on the young man's feet, at the end of the table where Ade and I stand, so that we can begin to open them. There is blood, of course, and it winds its way along the body's ankles and pools on the metal table. It also forms a low, still lake in his now-emptied abdomen and rib cage. The young man is pale, but his entire underside is mottled in reds and purples due to the hypostasis of blood. Without the heart's propulsion and distribution of blood—invisible to us, unwilled, unnoticed—gravity is the body's strongest force, and the blood sinks as low as it can, flooding the vessels and leaking out to color the tissues. Because he lies on his back, the purple seeps into his back, the undersides of his legs and arms, the heels of his feet, the back of his neck. It looks as if half of him has been terribly bruised.

Apart from all the other autopsies I ever see, the motorcyclist's body stands out because of the beautiful condition of his organs. When Ade cuts open his vessels, there are no dysmorphic yellow plaques of atherosclerosis, or "hardening of the arteries," found in every middle-aged and elderly body I see. His liver is smooth and dark, his kidneys shiny and perfect. I see them and think instantly of the comments a transplant surgeon made to a handful of my classmates and me during a small group-teaching session.

"Every state has a transplant list," he said. "Some waits are longer than others because of the number of patients waiting for organs, but some are longer just because there is a smaller supply of organs. If I have somebody who's really desperate," he continued, "I tell them to move to Florida." Everyone in the group looked quizzically at him. "I'm totally serious. If you're in a hurry for a transplant, you need to be in a warm-weather state where healthy boys ride too fast on their motorcycles year-round."

In the autopsy suite, Willy moves back to the kid's head and, with the scalpel, makes a cut over the top of it, beginning above one ear and ending at the top of the other. What happens next causes me to gasp. Willy slides his thumbs beneath the cut he has just made, beneath the mop of curly blond-brown hair, and peels the scalp down over the face, until the cut edge reaches the body's chin. He has essentially turned this face inside out, so that where the young man's features should be is instead the yellow-and-red underside of skin, hair curling oddly from beneath, at the level of his mouth. "Have yeh never seen the head done before, then?" Willy asks me in his thick Dublin accent, having apparently heard my sharp intake of breath.

"I have," I say. "I just forgot."

"Well, yeh can leave the room anytime yeh like," he says, smiling warmly.

"I'll be fine," I answer, grinning. "I'm tough."

He laughs. "So yeh are, then, luv. So yeh are." I have issued a knee-jerk response, a willful assurance that I can handle whatever medicine throws me.

The reality is that I am affected, profoundly so. As Ade is explaining his postmortem technique for examining the coronary arteries, I am focusing on the poor kid's hands, pale and cold, and on his feet, now bloodied by eviscerated organs, weighed and waiting to be cut apart and analyzed.

I have come to Ireland with Deborah, who is writing a comic play about an Irish art theft. Later that night in our tiny Dublin apartment when I see her, I find myself holding back the degree of my disturbance. She stops in the middle of typing to ask if anything is bothering me. I tell her that the commute to the hospital is tedious, that I will be happy to be done with the elective. I know that the truth is that I am not sleeping well, that the autopsies have brought back the sleep disturbances I had endured while dissecting Eve three years earlier.

I begin to weep in the kitchen before dinner one night a little while later, and I tell Deborah that it's awful how life is in someone one moment and gone the next, how I never, ever want that to happen to her. I can no longer tell how much of the emotion I'm feeling is about Deborah—and love and fear and mortality and me—and what is about a poor twenty-three-year-old who ruptured his spleen on his motorbike. *One minute he was standing up, talking to the medics,* Willy tells another technician who wanders through the room, *and the next minute he was dead.*

Just as a personal transformation must occur in order to quell the dissonance that dissection generates, a community transformation must occur for anatomical exploration to be embraced and encouraged. The prospect of a dissected body created—and continues to create—a great deal of confusion around ideas of mourning and resurrection. Though in America an increase in the prevalence of cremation testifies to the decreasing emphasis on the body in funerary proceedings, this shift is definitely not universal.

In Ireland, as elsewhere, the corpse, and the tradition of burial, is still a very central part of the mourning process. Those who have come to pay their respects to the family do so by filing past the body and shaking hands with the family beside the casket, sometimes stopping to touch or kiss the body. The tradition includes a graveside service, initiated by a procession of the casket from the church to the graveyard, followed by the whole congregation of mourners. The customary funeral is a weeklong affair, and the number of attendees at the funeral is considered to be a testimony to the life of the deceased. To eliminate the presence of the body—here, the true *embodiment* of the person whose life is being celebrated—disrupts this process. The result is a kind of cultural disorientation and unease, and it requires a redefinition of one of Ireland's most anchoring and fundamental societal rituals.

Clive Lee, the anatomy professor at the Royal College of Surgeons in Ireland, invites me to attend a biannual memorial service organized by the medical students, held to honor the people who have donated their bodies to the college. In order to express their gratitude to the bodies from which they have learned, students invite the friends and family members of the donors to gather in the Unitarian church on St. Stephen's Green. This year's service is held on a damp, cold January day.

I arrive a half hour early and take a seat in the corner of the sanctuary. Over the course of the next twenty minutes, a steady stream of people fill the pews. It is a crowd of all ages. Some people have come alone and others as large families. I am sitting behind a pair of elderly women. When six more women file in beside them in various states of religious dress, it becomes clear to me that they are all nuns and that many of them are simply not in traditional habits. When it comes time to sing hymns, I sing softly to hear their high, warbly voices that add harmonies and crescendos. They clearly know the songs and one another's voices with an intimacy that comes only from years of familiarity. I suspect that it is one of their sisters whom they have come here to honor, and I think of how far a distance they have traveled to arrive at this theological point from Boniface's edict and Vesalius' running to St. Anthony's basilica for mock absolution. Despite such religious history, the presence of these eight women makes much more logical sense to me. What more complete act of sacrifice and denial of attachment to the material world exists than to relinquish the vessel by which your soul was bound to it?

The nuns in attendance remind me of a phone conversation I had with my father after the autopsy of the young motorcyclist. Over the course of the conversation, he asked how my time at the hospital was going. I recounted to him, without going into much detail, that the process of the autopsy is quite difficult to watch.

"Is it just that the body is treated so brutally?" he asked. I paused for a moment to consider the question, then decided that it wasn't actually that at all. The process *is* brutal, I told him, but the upsetting part of the autopsy is not the way the body is handled, but rather that such handling makes no difference whatsoever. What one cannot quite comprehend, in the end, is that no matter what is done to the body, it has absolutely no effect on the person who once inhabited it. The horror is not what is present and cut apart but what has so completely and irreversibly gone.

The idea of a memorial such as this one is not altogether unusual. Many medical schools in the United States hold similar services, some just for the students to express their own appreciation and some, like this one, to which families and friends of the donors are invited. In a paper published in the *Anatomical Record*, Yale course director Lawrence Rizzolo states that the decision as to whether to invite donors' families is not necessarily a straightforward one. "Although the last several classes have considered inviting families of the donors," he writes, "none have chosen to do so." The ceremony, it seems, has become an institutional means of helping students cope appropriately with the extraordinary moments of doctoring. Rizzolo states that the concern has been that students would not feel comfortable fully expressing the "angst and frustration [associated with dissection] along with their joy and gratitude" if family members of the donors were present.

When Professor Lee steps to the podium of the Dublin church to offer his message to the people who have gathered for the Royal College of Surgeons service, he makes what I perceive to be an important distinction between this service and others by shifting the focus from the donors to the donors' families. In Ireland's body-centric mourning culture, this recognition is not just courteous, it is essential. Professor Lee expresses to the families that he understands and appreciates that they have undergone an unusually extended period of mourning. As is appropriate for a society in which bodily donation is so highly uncommon, he says he suspects that there are many in the audience who may not have known anyone else who has donated his or her body to medicine and that he hopes there is comfort in seeing all the other families here whose loved ones have also done so. Finally, and importantly, he thanks the family members for carrying out the wishes of the donors, "despite misgivings which you may have had" about their decisions.

Later I am sitting with Professor Lee in his office in the college, just down the street from the Unitarian church in which the memorial service was held. I tell him how happy I was to have seen the service and how nice I found his comments. He makes clear to me he felt that it was doubly important to address the families for reasons beyond the gratitude expressed by the students. First Professor Lee points out that an American medical school might get three hundred bodies bequeathed to it annually, whereas the number taken in by the Royal College of Surgeons in Ireland each year would be closer to twenty-five. For this reason, continued normalization of the process of donation and clear communication of the benefits conferred by such an altruistic act are an important investment in helping people consider donation as a viable and much-appreciated option. The second important reason, he explains, is a legal one.

"You can will me your jewelry, your furniture, your clothes, and your family won't have anything to say about it. You own those things. But, legally, we don't own our bodies," he says, gesturing with a kind of shrug meant to underscore the odd lack of logic. "This means that if you designate your body for donation and your son or daughter or spouse can't stomach the idea of it after you have died, they are under no legal obligation whatsoever to honor your decision." As he speaks, I begin to understand just how completely earnest, then, his words at the service were, thanking families for looking beyond their own misgivings in order to fulfill the specified wishes of their relatives.

Following the memorial service, all the attendees are invited to gather for refreshments in one of the rooms of the college. The idea is not only to continue to express appreciation to the families but also to allow them a chance to talk with the students whose careers have benefited from their loved ones' gifts. I did not attend, not wishing to disturb this fragile exchange, but as I filed from the church, I did see one interaction that must also be a benefit of the refreshment hour. Lines from several pews merged at the door, and as they did, two women in front of me recognized one another and said hello. An awkward pause followed, and then one of the women turned back to the other and said, "So it's both our mums, is it?" The other woman nodded. They both smiled and sighed in a way that seemed to acknowledge both the shared experience of losing a mother and the relief at having said out loud the fact of donation, which had perhaps been a truth at times too difficult to utter.

Mid-January darkness comes to Dublin late in the afternoon, and when I walk home from the church at nearly four o'clock, it is dusk already. On the pond in St. Stephen's Green, the mallards and a pair of tufted ducks tuck their beaks beneath their wings for sleep, ignoring the toddler throwing clumps of fist-mashed bread toward them with all her might. Behind me a bell rings in a repeated rhythm. It is a handbell held by the side of a limping man whose job it is to lock the park gates. He alerts us all that darkness approaches and we must leave. I stand for a moment longer to take in the last silver glint of the pond's oval, the floating ducks, the determined bread thrower in her orange coat and black patent-leather shoes. The air is cold and damp, but something about the canopy of trees in the green makes it feel as if every breath is purifying. The man is near me now, the bell ringing more loudly, a beautiful tone that sounds the uneven rhythm of his footfall. I take one last look, walk to the gates, and leave. I wonder if this is what death is like. A nighttime motorcycle ride, a glance at evening's light on water. A desire to stay to take in the beauty of the world with all its imperfections and a bell that sounds to say the hour has come beyond which we cannot possibly stay.

Just like the culture of detachment, which begins with legends of severed limbs and toll booths, the silence we all adopt marks another initial transformation on the road to becoming a doctor: the initiation into the culture of isolation. I am silent about my deep disturbance when Willy uncovers the young man's skull because I understand on some fundamental level that I am supposed to be. Here in medicine, because failure and weakness can cost people their lives, it is unacceptable to fail; it is unacceptable to be weak. Admission of either makes one seem unfit for the lofty charge. It is a culture currently blamed on litigiousness, but in truth its origins come from within medicine. Doctors stay up thirty-six hours at a time and subsist on vending-machine fare. They perform emergency surgery while others sleep. They maintain composure when the baby is lodged wrong in the birth canal, when the bone breaks through the skin, when the face is unrecognizably burned away. Part of this comes from necessity. But the problem arises when instead of setting aside our natural reactions, they are denied altogether. Then the culture simply becomes superhuman. And thus in the realm of the superhuman there is no room for human frailty, and admission of it by one risks revealing the illusion of the many. So no one speaks up, and as a result each person believes that she is alone in her experience. To that end, we are left in a profession with the pretense of untouchable greatness and infallibility, but one whose members kill themselves more than any other.

As medical students we are in the interspace between doctor and nondoctor. We have the spontaneous feelings that nondoctors do when we are first exposed to sickness and injury. We feel upset and disgusted and faint. Because we are in the interspace, however, we are aware that we must begin to manage those feelings and that, just as we must not blush when a woman opens her blouse nor cry when helping a family make decisions about a loved one's care, we must also touch and carry on conversations with patients who have seeping wounds, contagious rashes, disfiguring burns, bizarre delusions.

Allowing the feelings to wash over us in their fullness is no longer reasonable. Yet the idea of reaching the stage where we are unmoved by any of these things that so normally move other people is equally uncomfortable, if not more so. We understand how it can happen. After all, we are laughing now in anatomy lab. A paper in the medical journal *Clinical Anatomy* states bluntly, "Medical students' empathy tends to wane with each year of education, and by their third year many medical students want to distance themselves from their patients." How far are we from becoming the resident I follow on my surgery rotation who has had forty minutes of sleep for the umpteenth night in a row and, when a patient on another resident's service dies, says, "Why couldn't Mrs. Z have been under my coverage? Then I'd have one less patient to round on."

One night, eighteen hours into one of my twenty-four-hour shifts in the surgical intensive-care unit, a family had gathered around their patriarch, Mr. A. The patient's condition had steadily deteriorated over the course of weeks, and his body, having given out on practically every conceivable level, was being kept alive artificially—a roomful of large, state-of-the-art, whirring machines struggling to do the work that compact human organs had done effortlessly for decades. It had been weeks since Mr. A had shown any sign of consciousness, and there was no indication that he would do so in the future. One by one, the family members began to understand that their much-loved father and grandfather had no hope of improvement, and that the life remaining in him was borrowed from technology, that it no longer had origins in his body or his spirit. The family had gathered together at the bedside to discuss this grave reality, which had been confirmed by the doctors caring for the dying man.

Mr. A's physician, for whom a surgical resident and I were covering all night, had explained to the family that the ventilator, the dialysis, the endless medications could keep Mr. A alive for an indefinite amount of time in his present condition, but without any hope of improvement. The alternative would be to dial down the ventilator so that Mr. A received fewer breaths each minute. This shift in breathing would gradually mean that increasing amounts of carbon dioxide would accumulate in Mr. A's blood. Without adequate exhalations to expel what the body must normally breathe out, the failing body would be unable to compensate, and quietly, perhaps almost imperceptibly, in minutes or in hours, Mr. A would die.

After multiple grueling conversations, Mr. A's eldest son and daughter approached the resident and me. Their faces were dark with grief and drawn by the penetrating fatigue of not just this moment but of the past weeks. Calmly, they informed us that the entire family was in agreement: This kind of existence was not what Mr. A would ever have wanted. His loved ones were surrounding him—they would like for us please to turn down the ventilator, as their doctor had explained was possible. The resident agreed to do so. I nodded as the siblings spoke, in an attempt to support what had to have been a wrenching and awful decision for children to make about their parent. The conversation was a short one, and after it ended, the son wrapped his arm around his sister's shoulders and they headed back to the family, and their father, in the crowded room.

Once an hour for the next several hours, I quietly knocked and entered Mr. A's room to check on his exhausted family. Occasionally one or the other of them would leave for a tray of vending-machine coffee or to make a phone call to faraway relatives, but otherwise they remained. When four hours had passed with no apparent change in Mr. A's condition, I took the resident aside. "I know I haven't seen this kind of thing before," I said, "but are you surprised that Mr. A is still alive?"

"Not at all," he replied with a half smile. "In fact, I would have been surprised otherwise, since I haven't touched his vent since we came on this morning." I nearly stumbled as he spoke, picturing this family who was emptied of all reserves, awaiting their unenviable, but peaceful, finality of grief. My shock must have been easily recognizable, because the resident quickly continued, "Have you seen the paperwork I have to do for every death on this unit? His own doctor will be on in six hours and can dial the vent down then. No one dies on my clock."

We say we'll never get there. I say I'll never be that. But as doctors-in-training, we are reshaping the ways in which we react—in fact we are suppressing *universal* reactions of fear and grief and horror. Will I be able to suppress some but not all? Will I be able to detach from strangers and maintain the humanity with which I would respond to loved ones in similar circumstances? But then am I not bound to consider each patient as I would want my brother to be considered? My partner? My niece? The lines blur, and I am left feeling dissatisfied. I do not wish to blunt the spectrum of my feelings, to lose the discomfort I feel in violent movies, to lack empathy at the bedside of a dying patient. I do not wish to hear "stroke" and think of the distribution of vessels to the brain and the territories they serve instead of my grandmother's now-curled left hand and stooped walk. I do not wish to make love to my partner and think, *Latissimus dorsi, umbilicus, myocardium.* How much of becoming a doctor demands releasing the well-known and well-loved parts of my self?

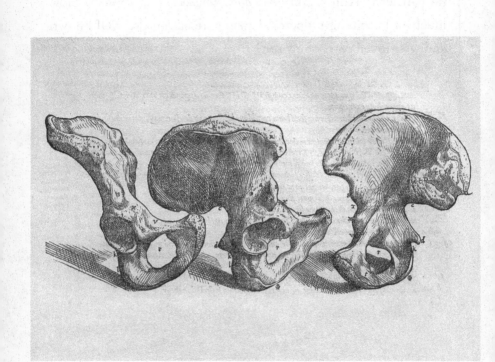

Pelvis

. . . mother

of my mother, old bone

tunnel through which I came.

you are sinking down into

your own veins, fingers

folding back into the hand,

day by day a slow retreat . . .

MARGARET ATWOOD, "FIVE POEMS FOR GRANDMOTHERS"

These initial forays into the body and disease inherently force my classmates and me to think about and define the boundaries of life and death. In strange ways the origin and the end of life sometimes overlap. One day in my final year of medical school, I am spending the afternoon on an inpatient hospice unit with Dr. Nancy Long, a wonderful end-of-life specialist. As we are rounding on patients, a dying woman's family approaches us. Of their loved one's final moments, they ask, full of grief, "How much longer can we expect this to go on?"

Dr. Long begins quietly, "Once someone starts to die, it is very much like when someone starts to be born. Just like labor, we know for certain that active dying means we are in the final stages, but we cannot say for certain when the end will come." Here she stops for a moment, until the family nods for her to continue. "A natural and human process has begun, and it has its own pace. We can only be here to participate in it and wait for it to progress." It seems right somehow that despite all the interventions we will learn over the course of our medical education, life is bracketed by forces that cannot ultimately be controlled.

For me, thoughts about life's boundaries emerge at both expected and unexpected times throughout medical school. Some of the most concentrated contemplation comes for me during the pelvic dissection in the seventh week of anatomy. I love the shape of the pelvis, as I have since the first day when I carried home my bone box and held that single pelvic bone in my hand. When we are learning the pelvic geography, Dr. Goslow brings a massive model of the bones to lecture, the size that an elephant might have. The model seems to me like a beautiful sculpture, with perfect, proportional, continuous lines and series of holes whose dimensions shift as the structure rotates: from one angle, mirror images of near circles separated by bone; from another, asymmetrical ovals; from another, creases of light, overlapping and oblong.

The language of these bones slides along their edges. Os coxae, the hip bones. Their three parts, with names like flowers: ilium, ischium, pubis. The place where all three parts fuse: acetabulum. Coccyx: ancestral vertebrae—a tail?—now fused. The pelvic brim, as if water spilled over it. The obturator foramen, the ischial tuberosity, the pubic arch, the ischial spine, the sciatic foramina. The names are so geographic it is impossible not to lose oneself in the massive model. Brim, arch, spine. The ligament names like a call to prayer: sacrospinous, sacrotuberous. Sacrosanct.

Vesalius's illustration of the bony pelvis and its accompanying annotation reflect an unusual lack of specificity when it comes to the nomenclature of these bones. Vesalius names the bone without naming it. That is, he calls it the innominate bone, or "the bone with no name." Galen had given no name to the bone. Vesalius, then, as a follower of Galen, subsequently refers to the bone as innominate, perhaps in deference to Galen's (conscious?) omission. Maybe it is this bone's inability to be encapsulated by language that makes me marvel. Or, equally probable, the fact that it is so unlike any other bone—or organic object—I know. Its effect is as when we see a lady slipper, and think, *Yes*, flower, *but utterly different from everything else that blooms*.

It may also be that there is something Platonic about this shape for me. Does the form resonate with me on some subconscious level, as if to say, *Motherhood, birth canal, womb?* Is my response to this shape different now, as I begin to feel a primal yearning for pregnancy? Or is this purely an echo of the unmistakable shape of the mothering from which I benefited? My young and beautiful mother, slim and sure of her body, wading backward in a hip-hugging two-piece swimsuit in the shallows of Higgins Lake on a northern Michigan summer day, gesturing for me to follow. How old am I in this memory? Two? Three? The orange swimsuit is what I fix my eyes on, determined to reach her. I am aimed toward reassurance and comfort. To her lap where I lay my head in the evenings after dinner. There she absentmindedly sorts through my tangled waves of still-damp hair and laughs with my handsome, larger-than-life father. Sleepy from swimming and sun, I follow the conversation around the room from my dad and mom to my New York artist aunt and uncle, my loud and wonderful grandmother, smoking her Winstons and retelling stories about the neighbors, my grandfather in light summer clothes with his sweet disposition and big, quiet smile.

My grandmother had this pelvic shape, too, bending over in her Indiana garden to hoe between the rows of peas, to dig potatoes, to let fat red and yellow tomatoes fall off the vine and into her hands. I can see her hips, strong and lean like a swimmer's, outlined in her white cotton pants as she lifts a bushel basket of apples into the station wagon or the tractor cart, and then again as she hoists tray upon tray of apples—Jonathan, Golden Delicious, Winesap—into the cider press. This shape of nurturing resonates. This outline of womanhood differentiated the bodies of my mother and grandmother from their husbands in their square swim trunks and khaki pants, from my brother's straight and lanky shape—all arms and legs, as if the torso were an afterthought—from my own thick child's form, which always seemed to me too big, too shapeless.

The pelvic dissection begins with the female genitalia. Nearly two months in the anatomy lab are behind us, but, regardless, this dissection is very strange to do. It feels so clinical and cold to peel back the labia with tweezers, to find the clitoris with a probe, to cut away the labia majora, exposing the labia minora and the vaginal opening whose numerous folds make it seem as if they cower in the light and turn in on themselves, so accustomed to being tucked away and shrouded. And whether it is because Eve's gender—and genitalia—are the same as my own or because of the interiority of the female genitals compared to the male, the probing and pulling feels particularly invasive. A probe does not belong in a woman's urethra. These are not parts that are meant to be touched by metal. In addition, it is an unnatural feeling to put a gloved finger into a stranger's vagina, and how much more a dead stranger?

When we are trained to do gynecological exams on the hospital wards, we will be told to tell the patients what we will be doing far before we do it: "You'll feel my finger at your vagina. I'm going to insert the speculum now. Some cold jelly and my two fingers going inside." The implication is that communication will decrease the discomfort of surprise, of unknowing. And also that the advance warning gives women time to respond: "The big speculum always hurts me. Can you use a smaller one?" or, "I hate this, so do it as quickly as you can," or even, "Once you've had children, you don't care what has to be done down there!" In any case, there is a dialogue. A seeking of permission, in essence, to enter into an intimate space.

Obviously, this is not possible with Eve, and I understand that she must have known that implicit in her agreement to have her body taken apart by students would be these moments of interiority. Still, the region is far more charged for us than was the relatively superficial forearm or even the distant, never-touched ventricles of the heart and knuckled shape of the spleen. The interiority of the genitals is associated with intimacy and in that way differs broadly from that of other parts of the body. These are spaces reserved for passion, for birth or the processes that allow it. This is not the neutral realm of the esophagus, the abdominal cavity, or the interior of the chest wall.

There are spaces in the body that lack the kind of emotional weight that the genitals possess but that share the idea of bodily entrance. At the mild end of the spectrum, we feel this when hooking a probe through the nostrils or deep into the back of the mouth. At the more extreme end, it is unnatural to push a probe through the eustachian tube of the ear, and it is an awful feeling to poke metal into the orifice of the eye. These are deeply held, practically innate lessons that even my toddler niece and nephew know—you don't put anything into these spaces in your body, and certainly not in anyone else's. You do not probe the ear, the socket of the eye, lest you cause damage. And we cannot help but feel as though we were.

The pelvic dissection does not get easier in the week after we have finished dissecting Eve's genitals. The summer before beginning medical school, I learned of the procedure performed in anatomy lab called the pelvic hemisection, in which the pelvis is literally sawed through at right angles so that one of the legs can be rolled to the side and removed. Since that time I have been dreading the dissection, and I do not now feel any less apprehensive about it. The night before the scheduled hemisection, I have no appetite and snap at Deborah for no reason. I sleep poorly, and when I do sleep, I clench my teeth so tightly that when I wake in the morning, my jaw aches and will not fully open. I try to tell myself that these are the physical manifestations of a mounting workload in all my medical-school classes, of the conceptual difficulty of the physiology of the kidney, of the seemingly impossible differentiation of tissue samples on slides in our histology course. What I know is that these are the physical manifestations of being asked to do something truly horrible to a body, to push beyond any shred of comfort or even interest. These are the sensations of breaking apart a previously inviolable internal wall.

The *Dissector*, in the end, does not pull punches. "Hemisection of the pelvis offers a unique opportunity to view many structures that lie deep within the pelvis and affords a chance to trace nerves and vessels," it begins. Then, frankly, "Check with your instructor to determine if you should complete this dissection."

The day is a difficult and emotional one. To begin, we pry Eve's legs apart as instructed, and put a rectangular wooden block between her ankles. We use the scalpel to make deep cuts through all the soft tissue we can, cutting into the body's side, above her iliac crest to the spine, and directly, inconceivably, down the middle of her vagina, clitoris, uterus, rectum, bladder. There is a different tone to the day for the whole room, and it is less tolerant of the lightness that has felt acceptable at other times.

When the soft-tissue cut is made, we are to take a ten-inch-long manual bone saw and saw through the spine crosswise and then up through the center of the pubic symphysis, the sacrum, and the lowest lumbar vertebrae (which means we must begin by sawing through the initial, lengthwise cut in the vagina and uterus). As we read the instructions in the *Dissector,* I feel dread in the pit of my stomach and remember the talk I have given myself that morning during my commute into school: *I was the one to saw the rib cage. I skinned her hand and cut off one of her labia majora. I don't have to saw her pelvis.* I know that Tripler won't volunteer to do it either, so I decide I will let Raj do it, or let him ask Trip or Tamara to.

I expect to feel shaken and upset once the sectioning actually begins. I do, partially. The sound of the bone saw is terrible, and we hear it, portentous, in other groups before we start the process ourselves. As is typical for them, Roxanne's tablemates are gung ho, and we are still cutting through the soft tissue when they have completely cut their man in two at the waist, are holding his legs in the air, with his low spine propped up on the table, and are sawing between the detached legs like mad. When the leg comes off, they let out a kind of cheer. The rest of the room lets out a deflated, instinctive groan of disapproval, and I feel sad for the spectacle of it. In contrast, our table's trepidation is evident. We wait and wait to begin, insisting that we need Dale or Dr. Goslow to help us find the barely visible bulbospongiosus fibers in Eve's innermost labia before we begin the process that will certainly render them unrecognizable.

Raj does begin the sawing, and when he does, I am holding Eve's right leg, pulling it slightly to the side. Trip is talking about going to visit her parents in Newport for the weekend, and I focus as much as I can on her mouth moving and her chirpy, high-pitched voice. The sawing goes on for a very long time—longer than anyone would have expected—and we stop several times to assess our progress, to break for lunch, to see a prosection. Raj and I are alone to do the final push of it, and he grows tired from all the sawing just before the end, so I reluctantly take over, sawing into Eve's vertebrae. It is my cut that causes her leg to roll to the side, exposing her split, apricot-size uterus and its thick inner wall. She is now in two parts—her amputated leg and the body from which it came.

This dissection causes a kind of shift in the emotional momentum of the course for my lab partners and me. Where we had been growing increasingly at ease with the actions we performed upon Eve, we now grow less so. And where her partitioned body had begun to resemble a human being less and less, it now harbors new and undeniable reminders—like this small fist of uterus—of the living body that Eve once was. Each cut we make into Eve's flesh should reiterate her lifelessness, and yet the distinction between "alive" and "dead" is not always as evident as it rationally should be.

My classmates and I are not the first to stumble into the netherworld of these boundaries between life and death. Countless cultural rituals and superstitions surround the liminal period between death and burial. In many modern and historical cultures, there exists a widespread belief in a sort of liminal state of consciousness immediately following death. The duration of this period differs significantly—in some instances just until the body's temperature becomes cold, in others until some spiritual quest is resolved or reconciliation occurs; in many there is a sense of ongoing personhood or consciousness that is permanent.

Ancient Egyptians, for instance, performed a ceremony in which the mouths of the dead were pried open so that the deceased could speak important passwords in the afterlife. Still today, in rural China, the parents of young men who die as bachelors search for recently deceased unmarried women to whom their sons can be "married" in death. The couples are interred together in the hope that they will have happy afterlives. The wealthiest of these families may slaughter an animal and have a reception to honor the union; those too poor to afford even a dead woman's dowry may bury their sons beside bridal figures made of straw.

In nineteenth-century Britain, it was commonly thought that the unburied corpse might return to life. In recognition of this possibility, food and wine were left by the body so that hunger and thirst would not go unsatisfied if the corpse should awaken. Even if the body was to remain dead, it was commonly believed to be able to defend its wishes, and help the cause of justice in the case of wrongdoing. Many traditions state that the corpse was capable of showing displeasure—a murder victim would speak or begin to bleed if the murderer walked into the room, as would the deceased person whose falsified will was read in his or her presence. The most dubious law surrounding the newly dead established their ability to enter into agreements and sign documents—any mark or signature taken from the hand of the corpse was held in the same legal regard as one made before death, so long as the mark was made while the body was still warm.

As is true for so many beliefs and superstitions, it is likely that these initially arose from real circumstances: remnants of blood that happened to come into view when the body was shifted by an unlucky mourner's touch, a last release of air from the mouth as the muscles of respiration began to collapse. And if we, with our broad base of knowledge and highly sensitive medical equipment, at times struggle with the definitions of life and death, it is not surprising that it would be murky territory for those who relied upon the presence or absence of the fog of breath on a mirror to delineate between the two. Indeed, the popular lack of confidence in the ability to resolutely determine death led to fears of being buried alive. Hans Christian Andersen, for instance, is said to have propped a note up on his nightstand each night before sleep that read, "*I am not dead.*"

Strange superstitions also surrounded the bodies of executed criminals during the nineteenth century. It is tempting to think that the superstitions arose from a sentiment that I, too, have had: an inability to fully believe that once dead, the people we have known lack any powers or abilities at all. Yet the specific power that criminal bodies seemed to possess may, in fact, have indicated a belief in the powers of dark forces rather than a disbelief in the powers of death. Both of the two most commonly held superstitions involved the criminals' hands but were used to very different effect.

The "Dead Hand" was technically the severed hand of any corpse, thought to possess healing powers. The most powerful Dead Hands, however, were those that had once belonged to newly executed criminals or suicide victims. The Dead Hand was used to rub ailing body parts and was particularly helpful for sufferers of cancer, tuberculosis, warts and other sores, but it was also said to help with neck and throat problems and (though the treatment conjures disturbing images) female infertility! Incongruously, the severed hand was also believed to speed the churning of butter when used to stir the milk.

The "Hand of Glory," conversely, despite its rather religious-sounding name, was used by witches and thieves throughout Europe. The hand of a hanged felon was severed from a body still on the gallows, pickled and dried. A candle, made in part from the fat of a hanged man, was fixed between the fingers. When lighted, such a candle would cause occupants of a home to fall into a deep slumber, uninterruptible by even the greatest noise, allowing thieves to work in the house at will. Some variations of the Hand of Glory lighted the fingers of the hand itself, rather than a mounted candle. If the finger would not stay lit, it was said to indicate that someone in the house remained awake. Perhaps this omen was ignored by the burglars who were discovered using just such a Hand of Glory in a foiled robbery attempt in rural Ireland as recently as 1831.

Various spiritual beliefs shaped ideas about the afterlife. In a wonderful Norse tale, ghosts haunt an Icelandic castle. The ghosts take claim of the fireplaced rooms and force the owners to the cold and gloomy recesses of their home. The owners are intimidated by their unwanted houseguests but eventually are driven to try to reclaim their home. In order to do so, they air their official grievances at a full trial by jury. There the ghosts, who are present, are convicted and sentenced to permanent exile from the castle. Apparently having no choice other than to comply, they do so, thereby leaving the legally ordained, rightful owners in peace.

Pagan as these beliefs may sound, examples abound of religious authority relying on just such principles. Pope Stephen VI, in the ninth century, claimed that his papal predecessor, Formosus, had been disloyal to the church. Stephen actually took the corpse of Formosus to trial, where, after the appropriate proceedings, it was promptly pronounced guilty. As punishment, the right hand was severed from the dead body of the accused, mutilated, and finally, for good measure, thrown into the Tiber.

Though this anecdote of long-ago ecclesiastical judgment is an outrageous one, it does not stand alone as an illustration of the church's participation in the bodily interspace between life and death. Deborah and I found a vivid example of just this tension while I was researching anatomical history in Italy: the incorrupt corpse of Santa Caterina of Bologna.

St. Catherine's relics are housed in a small church called Corpus Domini, and when we arrived, the chapel of the saint was closed, the church empty. We rang the buzzer of the convent beside the church and waited a long while. Nothing happened. When we were considering whether we might try to find someone within the church, a female voice came from a tiny speaker above the button: "*Si?*"

I responded in terribly broken Italian: "*Buongiorno*. Sorry. No. Speak. Italian. To see. Saint?" The voice came back over the speaker, and I understood her to say that the saint's chapel was closed. We could come back tomorrow. I tried again: "Leave. Today. Is possible. To see. Saint?"

A long silence. Then in English: "Open door, please." When I did, no one was there.

"Over there," Deborah whispered, nodding in the direction of a white wall, which on closer inspection had a white mesh square in the middle, like a confessional screen. Behind it a barely visible dark shape moved.

As we approached, we heard, "Go into church. I will open door in five minutes."

"*Grazie, grazie!*" I said, leaning toward the screen, but it was now all white, and any trace of movement had gone.

In the silence of the church, the lock turning in the door sounded like a shot, and we jumped at the noise. When I opened the door, there was no sign of the nun who must have been there only seconds before. With no other option but to go forward, we began walking. The door closed solidly behind us. At the end of a hallway was a very small chapel, glittering with gold, its ceiling ornately painted in pastels. And there, in the center of the room, seated upright in a glass chamber canopied in red-and-gold brocade, was the 541-year-old corpse of Santa Caterina of Bologna.

Unlike countless saints whose bodies are scattered in pieces between tens or even hundreds of churches, St. Catherine's body is completely intact. She wears a traditional habit, and her bare toes peek out from beneath its hem. In the center of the chapel wall directly across from Catherine is a porthole through which the saint's face—a wizened dark mahogany hue like her hands and toes—can be seen from the sanctuary of the church.

Reliquaries from other saints surround Catherine. A skull and a pile of bones rest on a shelf. A large glass box holds a board with mounted rows of teeth—too many to have come from one mouth. Numerous small boxes contain the staples of holy containers found in nearly every church in Italy: vials of saintly blood, bits of skin and fingernails, the small bones of the wrist, a desiccated toe. It is impossible to know how many saints are represented here, their fragments positioned to pay homage to the upright corpse. Catherine's head is cocked a little, as if she were listening to something that the bits of saints might say.

Deborah and I were alone in the chapel with a person who died in 1463. The nun whose shadowy presence led us here was nowhere to be seen. We tried to behave piously, while at the same time desperate to glance at each other in the total silence to say, *Can you believe this?* Two offering boxes were positioned at Catherine's feet, along with two benches upon which we knelt until a sudden *whoosh* to my right nearly made me jump out of my skin. It was the disappearing nun, who slid open a fabric window through which she must have watched us arrive. I stood to thank her. She did not meet my eyes but wordlessly slipped something through the window, closed the fabric, and was gone. I picked up what she left us: two tiny prayer cards in English. On the front, in glossy color, was a photograph of the dead saint on her golden throne.

Catherine of Bologna was an accomplished miniaturist and was therefore named the patron saint of artists. In death she is said to have appeared to a seventeenth-century nun with proposed designs for the ornate chapel where Deborah and I knelt, so that her incorrupt body could be properly enshrined. As I studied her withered face, I thought that Catherine's glass-encased corpse was the perfect symbol for the Renaissance collision of art and science and religion and death. In a faith that forbade dissection, Catherine would be the patron saint of Leonardo in nearby Vinci, eleven years old when Catherine died, or Michelangelo, born twelve years after her death—two artists who zealously attended dissections, even took legs or heads or bodies to their own studios for study. In an era that condemned the fact that dissection denied the dead the burial necessary for the soul's rest, resurrection, and eternal life, Catherine's sanctified corpse was removed from the grave (and placed in a not particularly rest-inducing position) with no intention of reburial. And in countless churches, the enshrined bits and pieces of the most religiously revered human bodies in the Catholic faith are displayed, partitioned and segmented, cleaved and severed and torn. If there is a difference between the physical treatment of these saints' bodies and that of the bodies of the dissected, which were deemed desecrated, I cannot see it. And, I suspect, neither could Vesalius, kneeling before the tongue and larynx of St. Anthony of Padua.

When Deborah and I left the holy corpse and reentered the sanctuary, we stood in front of the round window that looked into the chapel. Santa Caterina's face was framed in a circle of light, and though we could not see any more of her body, or the room, a shimmer of gold encircled her. After only a moment, the light snapped off. The ghostly nun must have known we left and, once we did, turned out the chapel lights. When we peered into the circle, we could now see only the round glass of the window, a framed lens of darkness in which our own reflected faces were visible.

Though her purportedly incorruptible flesh and her ornate chapel give St. Catherine a relatively unique position among relics, the idea of a holy body on display rather than buried is certainly not an anomaly. One of the most vivid, and currently accessible, examples of the church's invitation to the living to view the holy dead is the crypt of Santa Maria della Concezione, in the heart of downtown Rome. The church's front looks like any other among the hundreds that dot the Roman streets and alleyways, and even the presence of a crypt in any such church is more the rule than the exception. In fact, the only clue that even hints at the oddities behind the doors of Santa Maria della Concezione is the name of the street leading to the church: Via dei Cappuccini—the Street of the Capuchins.

Beneath the sanctuary of Santa Maria della Concezione, the bones of four thousand Capuchin friars have been arranged in a series of chapels that, even after my time in anatomical museums and conducting cadaveric dissection, remain the most bizarre thing I have seen. The skeletons of monks, still clothed in their dark brown tunics, stand hunched in arched spaces. Between and over them are the meticulously stacked bones of their brethren, their leg and arm bones, their jawless skulls.

Without question it is not the bodies that unnerve the most, or even the piles of bones that, if one begins to count how many lives are represented by each sacrum or jawbone, cause the mind to spin and reel. The most disconcerting element of the crypt is that the bones are used not just as memorials or relics but omnipresently as décor, which is injected with a creepy kind of whimsy. From the ceilings hang illuminated chandeliers of cantilevered arm bones and clavicles. The doorways between the chapels are outlined by a filigree of looping ribs. Half pelvises on the walls and ceilings overlap at a central point, flowering out in petaled whorls, and vertebrae are gathered in clusters, forming oddly beautiful wreaths.

In some spots the collages of bone convey portentous, *tempus fugit*-style messages of mortality: The pointed base of one sacrum abuts the base of another so that they, outlined by matching clavicles, form the shape of an hourglass. Two scapulae branch out from the sides of a skull, like wings. On one ceiling the skeleton of a child is mounted—one of three nephews of Pope Urban VIII on display in the crypts. In one small hand, he holds a grim reaper's scythe. In the other the scales of justice, whose chains are made of finger bones, interlaced and dangling.

There is little explanation of why the crypt is as it is. Informational signs explain that in 1631 the remains of four thousand friars were exhumed, but nothing addresses the question of why the logical thing to do with them was to turn them into bony lamps and flowers. It is the shift away from the solemn idolatry of reliquaries that makes the crypt feel like the playground of some lone sociopathic monk, until you read further and realize that the crypt was maintained and added to—presumably as a now-incomprehensible act of religious devotion—from 1631 until 1870.

The ongoing decorative arrangement of Capuchin skeletons—and the seventeenth-century construction of the gilded chapel and throne for Santa Caterina—neatly overlapped with the religious and public outrage over anatomical dissection, underscoring a paradoxical—and perhaps evolving—view of the dead body and its importance. Indeed, to deny a body eternal burial was sacrilege. However, whether in medicine or religion, the boundaries of life were beginning to extend, granting dead bodies a role of importance and a chance to improve the lives of those still living.

When we are instructed to split the pelvises of our cadavers in two, it necessarily requires an additional distancing from the life, from the personhood of the bodies. *Detached concern*. Perhaps because of this, I am, in the end, very tired, but less bothered by the hemisection than I had imagined I would be. The reactions to the day from my classmates, however, are deeply varied. Helen, a warm and peppy twenty-two-year-old, begins to cry during lab. Many people seem testy, angry at jokes their tablemates have made. Everyone understands that the jokes are merely another way to deal with the emotion of the day, but one that in this instance can more easily bleed into presumed disrespect. Everyone looks a bit haggard and seems to wish the day over much sooner than normal. At the end of the day, I think to myself that I have done one of the strangest things I will ever have to do.

The uterus is amazing. The fallopian tubes are obvious, and the tiny fimbria look like a baseball mitt ready to catch the perfectly aimed ovum. I picture a balance that is this day, this pelvic hemisection. One side is weighed down on the balance's tray by Helen's tears, by the harsh words I spoke to Deborah the night before the scheduled lab, by the time and the amount of force required to saw between a body's legs, by the sounds that a joint makes as its connections give and break.

On the other side lies wonder: this small, perfect swell of an elderly woman's uterus. What a surprise to see in reality the picture of my fertility. Here are the small blue ovaries that cannot be described as anything other than egg-shaped. Here are the fimbria, from the Latin meaning "fringe," here the fallopian tubes, small, but sturdy and purposeful. And here the uterus; how to imagine that this tiny, misshapen ball stretches and grows to hold a new person within it? How to make sense of its capacity to fiercely and forcefully contract, to push a baby-size baby out of the small tunnel that is its entrance?

It is this moment that makes me wish to speak to Eve. I do not know which weight the balance favors. I do not know whether this new acquisition of knowledge, this vision of her depths is worth what I have done to her. I do not know if it is worth it to me. But from Eve, with her leg sawed off and rolled aside, with her chest now open, her bowels in a bag beneath her, what I wish to know is the story of her uterus. I wish I could ask her: *Children, Eve? Joy? Heartbreak? Catastrophe? Passion? Power?* I see this small space of hers and wonder how, if at all, it was tied to who she was, to the life she led. I wonder if her bodily womanhood was centered here or someplace else, more subtle, that I would never guess. Nonetheless, I reach into her to touch this space she never saw and wish I could offer it a blessing.

At first, in our training as doctors-to-be, we are given medicine at its most clear-cut. Thankfully—and necessarily—paired with the feelings of transgression we experience during dissections are teaching sessions that are objective, factual, and emotionless. Here is the context that will allow us to think of these elements of the body as functioning parts to a systemic whole. These are the gears of the machine that we may be called upon to find and to fix. So think like a mechanic and pay attention. In general, this works well. We are taught the way the genitals develop in men and women from shared fetal anatomic precursors; this accounts for findings like the small slips of muscle around the labia minora that are elusive, but present, and parallel the more evident muscular structures in the male. In this, our first exposure to the fork in the developmental road of gender, we learn how each fetus, regardless of chromosomal gender, is equipped with the physical tissue for both testes and ovaries and for both male and female external genital structures. We learn that a cascade of hormones causes either the male or female genitalia to take root and grow and, consequently, to cause the opposite gender's structures to wither and fade.

What we do not learn now, but what we will learn later, is that, like all bodily pathways, there can be wrong turns, missed steps, interrupted directions. We learn that even physical gender—one of the physiologic distinctions we take as the most basic—is not nearly the black-or-white, male-or-female, pink-or-blue differentiation we have classified it to be. In the back of my mind somewhere during this lecture is the disturbed awe I felt while reading John Colapinto's mind-boggling *As Nature Made Him: The Boy Who Was Raised as a Girl*—that there are some for whom this developmental fork is not distinct. *Robbins' Pathologic Basis of Disease*, the most widely used pathology textbook in medical schools, acknowledges the murkiness of this territory:

> The problem of sexual ambiguity is exceedingly complex. . . . It will be no surprise to medical students that the sex of an individual can be defined on several levels. *Genetic sex* is determined by the presence or absence of a Y chromosome. No matter how many X chromosomes are present, a single Y chromosome dictates testicular development and the genetic male gender. . . . *Gonadal sex* is based on the histologic characteristics of the gonads. *Ductal sex* depends on the presence of derivatives of the müllerian or wolffian ducts. *Phenotypic*, or *genital*, sex is based on

the appearance of the external genitalia. Sexual ambiguity is present whenever there is disagreement among these various criteria for determining sex.

This lack of clear definition—even lack of clarity regarding the criteria for definition—surfaces and resurfaces as a theme in medicine. What is male and what female? When is a person alive, and when dead? At times, in fact at *most* times, specific knowledge in medicine seems to be better understood than general knowledge. We might know the exact, microscopic ways in which hormones and chromosomes alter embryological tissue to yield a child with both ovarian and testicular tissue, but we cannot with confidence say whether this child is a baby boy or a baby girl. The realization that looms is that the categories upon which we build our understandings of humanity do not leave room for those who fall between them. And as shifting beliefs and medical advances mean that a baby with both a vagina and a penis can no longer be left to die in the wilderness or relegated to circus sideshows or medical exploitation, those who are not easily categorized also do not easily disappear.

When, during my third year of medical school, I am rotating through the surgical intensive-care unit and working twenty-four-hour shifts every other day, I witness firsthand the inexactness of medically defined states—in this case alive versus dead. Mr. W has been in intensive care for two weeks; his skin is totally yellow. His testicles are the size of grapefruits. Mr. W no longer responds to painful stimuli.

"What does that mean?" Deborah asks when I am recounting

to her the upsetting condition of Mr. W. I tell her that it means every morning, in our assessment of his condition, we take our thumbnails and push down, hard, in his fingernail beds, and that he doesn't pull his hands away. I tell her that it means we make a fist and press our knuckles, hard, into Mr. W's breastbone, and he doesn't respond. It is the standard way in which the most basic neurological reflex—retraction from pain—is measured, and we do so daily. We pull his always-closed eyelids open and shine our penlights in to see whether his pupils constrict.

"Maybe a little," says the resident on duty. The attending physician shrugs. I take my light from my white coat pocket, pull his left lid up with my gloved thumb, shine the beam on his pupil, and see nothing. I aim the light away, then back over his pupil. Still nothing. I know that he is not conscious, that we are not even sure how *alive* he is, that I can't see any change in his pupils with light, but he still seems to be staring straight at me. Unflinching. I take my thumb away, allow his eye to close.

"It's subtle," says the resident.

"So how do you define someone as dead?" Deborah asks when I tell her about Mr. W. I don't have an answer. We talk about it for hours, even though my schedule has me so tired that I can't come up with the words I want to use, that I have to ask her to repeat herself, because unpredictably, and unintentionally, my concentration drifts away in the middle of her sentences, and I am thinking about the pattern on our napkins or the twitch in my eyelid that comes on every few hours and has during my whole tired surgical rotation.

This conversation is one we'll repeat throughout medical school. It is one I raise when I care for a fifty-six-year-old man who has been brought to the hospital by ambulance after having been "down in the field for a long time." The translation is that he had a heart attack and was resuscitated. But in between those two happenings, enough time passed without his brain getting adequate oxygen that it left him unable to communicate, unresponsive, and yet making disconcerting, patterned movements. The neurologists call them primitive gestures, meaning that they are indicative of nothing purposeful, meaning that his brain has been reduced to putting out meaningless firings of neurons, which meaninglessly move muscles, cause his head to twist to the side, his elbows and wrists to flex, his lips to smack and tongue to protrude. For doctors these symptoms are often a clear indication of the appropriately named cerebral devastation, and a hopeless prognosis. For families these are movements that *must* offer promise, mustn't they? *If we're supposed to believe he's brain dead, then why are his eyes open? Why is he moving?*

"It's okay, honey, we know you're in there. You don't have to convince us that you're still fighting. We can see it plain as day." The words spoken by desperate family members are pointed, directed not at their loved one but at us, his medical providers, as if to say, *Don't you try to talk him out of a chance.* Are we? The truth is that this is an instance when medicine seems to fly in the face of reason. The other side of the truth is that for every hundred, or thousand, or even ten thousand Mr. W's who have these symptoms and waste away and never recover, there is the one miraculous story of a woman who comes back from certain brain death, from months or years in a coma, from, the patient's family says, *these exact symptoms*. We health-care workers hear this and give a knowing (patronizing?) smile, clear our throats, explain that those instances were probably examples of misdiagnosis, of inadequate testing, that they were probably different and less clear-cut than this scenario.

And even if they were not . . . And here is the rough truth of it, where the voice trails off. The end of the sentence is this: Even if they were exactly the same as this, and through some miracle of miracles, recovery was made, is it necessarily morally right to keep vigil? To spend hundreds of thousands of dollars on a gamble with odds like the lottery? To allocate medical resources and, dare I say it, perfectly good organs, which could be, if taken right now, nearly guaranteed to give a *good* chance to someone else, to this long shot of long shots? The answer, almost always, is: Easy for us to say. The answer is that that girl who needs a liver is not my son. My son is here, having just had a terrible motorcycle crash, or a heart attack, or massive surgical complications, and I am on the brink of losing him. The answer is that the utilitarian perspective of resource allocation and odds is a luxury afforded to someone who does not love the body lying in the hospital bed with oddly flexed arms. The answer is that almost no patient's family can know—can bear to know—just how slim the odds are, let alone acknowledge that and make an irreversible decision based upon it.

These are the kinds of questions I share with Deborah after dinner, after the lights go out. What does it mean to be dead?

How, short of the rigid, wrong-colored, breathless, embalmed state of Eve, to *define* dead from alive? These are the gnawing questions that *must* be shared in some capacity, or the impossibility of their answers creates a space between doctor and nondoctor that, in any loving relationship, is difficult to reconcile.

And yet I absolutely do not tell Deborah everything. The distinction is wholly mine, completely arbitrary, and I cannot know how much of it is truly a valiant attempt not to disturb her and how much is my own need not to revisit the things that upset me most or, as is sometimes the case, upset me most inexplicably. When I first see a laparoscopic endometrial ablation to stop a woman's dysfunctional uterine bleeding, I tell Deborah that the inside of the uterus looks soft and pink and red like cotton candy, but I do not tell her that to complete the procedure the resident rolls an electrified metal wheel along all the surfaces, burning out the pink and red in strips until the whole surface is white, tinged with brown char, lunar.

I do not know for certain the first time I see someone die. Which one you count depends on definition, on timing.

In February of my third year, I will run to a room in the veterans' hospital when the overhead announces, *Code Blue, Room 576, Code Blue, Room 576.* By the time I arrive from the basement cafeteria, the room is full. Nurses and doctors have run from all over the hospital in response to the announcement. A resident stands in the corner of the room on a chair, shouting instructions.

Nurses fill syringes and hand commanded items to other residents: *I need another milligram of epinephrine, a new pair of gloves, where the hell is that atropine I need* now? A medical student does chest compressions on a pale, lifeless, naked man. *One and two and three and four and exhale* (breath from strong and living twenty-four-year-old man inflating the lungs of dying seventy-three-year-old man), *breathe, exhale again.* Every few minutes the resident in the corner shouts, *Stop compressions!* and all eyes watch the EKG monitor to see whether there are beats on the screen. At other codes I will have seen, there were. This time, and all subsequent times this day, for this man, there will not be. The cycle resumes several times. CPR. Epinephrine. Atropine. *Stop compressions!* A flat, green line on the monitor.

Was he dead when I arrived? Or dead when the resident on the chair said, "Does anyone have any objections if I call this? Okay. Eleven twenty-nine. Thank you, everybody."

After the man died, an intern said to me, "You should have gotten in there to do compressions so you know what the ribs feel like when they break. Do you want to try it on him now?"

W hich one you count depends on definition, on timing.

It is April of my third year, two months after the code blue at the VA. He is a new patient to hospice that afternoon, and therefore I am assigned to gather his history, do a physical exam. This is not new to me. I have been to hospice one day a week now for four months, and I know what dying people look like. Once they have been sick long enough, as most who are in hospice have, they grow thin. They are pale. When they sleep, which can be all the time, their mouths are open wide. This has its own terminology outside hospice, which was explained to me by a surgery resident. "That's the O sign, considered a poor prognostic factor, but not as bad as the Q sign," he said, grinning, referring to the open mouth with the tongue lolling out of whichever corner is most susceptible to gravity.

When death is imminent, the skin mottles. It looks as if reddish blue yarn has been crocheted beneath the skin. The hands and feet typically grow unnaturally cold. Both of these changes are indications of circulatory failure.

Occasionally, paradoxically, the extremities feel as if they are on fire. A last, feverish flare. This I felt once, on the hands of a Portuguese woman who was dying alone, a portrait of Jesus tacked above the head of her bed.

When I enter the room to gather information from the man who has just arrived, I cannot do so. He has amyotrophic lateral sclerosis, also called Lou Gehrig's disease. He cannot speak or gesture. In fact, he can move nothing but his eyes. One blink of the left eyelid means yes, the nurse tells me. Two blinks mean no. These are the only movements he can make. Nearly everything else is paralyzed—his arms and legs long ago, his trunk afterward. Then his vocal cords. Only this weekend his ability to swallow. I know that when his diaphragm succumbs, he will suffocate and die. His breathing is becoming progressively more difficult, she tells me. He is forty years old.

He stares at me, unwavering. I introduce myself. I obviously forgo the history and settle for essential information.

Mr. L, are you uncomfortable?

Blink. *Yes.*

Are you in pain?

Blink. Blink. *No.*

Is it still hard to breathe?

Blink. *Yes.*

Is the medicine they gave you to help that making any difference?

Blink. Blink. *No.*

I'll be right back.

I walk quickly to find the doctor, and she prepares to give him more medication—morphine to help his shortness of breath and some Ativan, an antianxiety drug, to help calm him. When I ask her why Ativan, she says, "Read this," and hands me Mr. L's chart. In a note written by his primary-care physician many months ago, is written, "Mr. L continues to have severe and distinct fear in regards to the potential loss of respiratory function, as has been his primary fear since the day of his diagnosis."

When I return to his room, I tell him that more medicine is coming and reassure him that it should make his breathing easier. He lies atop his covers, wearing only the adult version of a diaper. He is emaciated. His skeletal feet turn toward one another so that the toes of his left foot overlap those of his right. He shows the textbook signs of respiratory distress. When he breathes, every conceivable muscle seems to try to help him do so. The space beneath the sternum retracts. The spaces between the ribs invert. Even his nostrils are flared to allow as much air to enter as possible.

Are you feeling scared, Mr. L?

Blink. *Yes.* It is emphatic. And so is the stare that follows.

I tell him that the Ativan is coming, too, in addition to more of the first medicine, that it will help his anxiety. I cannot know whether I am lying. How strong a medication does one need to take away the feeling of terror each of us feels when we cannot get enough air into our lungs? It is felt in asthma attacks. It is felt at high altitudes. With too much exertion. It is felt when we are caught in currents underwater. When we inhale water and cannot stop choking. How much worse when paired with the knowledge that it is a symbol for the end of an irreversible process? How much more do we fear the symptom when we know what the symptom implies?

I do not tell him that I don't know if the Ativan will touch the fear he feels. And I do not tell him that his is the disease I fear the most of every one I have studied. He is the embodiment of what I never, ever want to experience. What I never want a loved one to know. Give me cancer, if you must. Give me Alzheimer's. Give me emphysema, congestive heart failure, diabetes. It is this disease that I fear each time my eyelid twitches, each time I feel numbness in an unexpected place. Is it . . . ? Could it possibly be . . . ? Mr. L, please do not ever let me be you.

His parents arrive and, unbidden, deliver the history that I was originally seeking. Mr. L went to the doctor's two years ago with a limp that wasn't going away. It was interfering with his rock climbing. Mr. L was an adventurer who worked to support his travels. His home was decorated with maps. His shelves lined with guidebooks. When the limp worsened despite physical therapy, the doctor did further testing. There was no familial history of neurological disease. Mr. L had had no previous significant illnesses. He had no symptoms, other than the limp. With the diagnosis came gradual decline. Mr. L moved back in with his parents and eventually became immobile. This weekend when they fed him, they noticed that the bites of food never left his mouth to travel down his throat. When he began to struggle to breathe, they took him to the hospital, assuming (probably correctly) that he had aspirated some of his food and now had pneumonia, decreasing his ability to breathe fully and enough. Mr. L had designated in a living will that he did not wish to be intubated, to have a machine breathe for him, thus potentially prolonging his life significantly. But with eventual paralysis of his eyelids and then no ability to interact with others at all, despite a fully conscious interior mind, the hospital could do nothing further for him and so transferred him to hospice, where his comfort would be the primary medical goal.

I leave his parents for them to spend time with him and to hopefully help him feel calmer. Moments later his mother rushes up to me, and looks scared. "He's having a terrible time breathing, and he looks awful," she says. I find the doctor immediately, and we go to Mr. L's room, where his father is standing by him, helplessly looking for us to arrive. The hospice nurse is there, answering the parents' questions, arranging pillows, using a suction device to clear secretions from Mr. L's mouth when his breath sounds coarse and rattling.

Mr. L's eyes are open wide, and he is gasping big, irregular breaths. He is a foreboding shade of gray. And though his eyes stare straight at me, it feels to me as if it were only because I am at the foot of his bed, and that is where his eyes happen to be directed. It feels as if he were absent from the gaze. Or perhaps I cannot bear, cannot understand the thought of his trying to communicate something to me in this moment. His parents look to the doctor, who calmly, gently, explains to them that the raking sounds of his breaths sound worse to us than they feel to him due to the medication we have given him and, though he looks distressed, it is unlikely that he feels so. The parents ask what is happening. The nurse says that we never know, but that it seems as if he may be taking his last breaths. The doctor nods in agreement. I am dumbfounded.

The parents each let out their own, independent, primal moans. It is as if this chorus were a parental reflex—what you do when you are told that you are watching your child die. Yet, unlike children at the bedsides of their dying parents, these two do not have any kind of innate sense of what to do. The moment defies the natural order. Where is the innate response of action in a parent whose child is dying? Perhaps there is none.

The nurse guides them. Tells them that he can hear them. Can he? Regardless, it is better for everyone if we believe it is so. She tells them to take his hands if they wish. They do. The father falls to his knees and sobs. The mother collapses into a chair that the nurse slides beneath her. I cannot look at the father without beginning to cry. I look straight back into the eyes of Mr. L, which look straight back at me. I grow increasingly convinced of his absence behind his gaze—otherwise why would he fix it on me? Why, with his parents beside him, with a doctor beside me? Nonetheless, I try to smile a comforting smile, in case he can see as well as hear. I try to convey to him peace. I expect that I am not particularly convincing, though in the moment I believe that I can be. But I do not feel peace inside. What I feel is emotion, and then a sense that I am not entitled to the emotion. I feel sadness, yet who am I to feel sadness when I am in the presence of parents watching their son die? I feel fear, yet how could it compare to the fear of this young man, taking labored breaths that propel him toward death's uncertainty? I wish him to live, but I am sure that is only to relieve my own discomfort, and, if Mr. L's living will is any indication, it is not necessarily what he desires. As minutes pass and his breathing seems to grow more and more pained, as his parents seem less and less sure what they should be saying, which breath will be his last, I begin to wish he would die. To end this fraught stretch. I do not know if it is a selfish wish. It is an interminable time. I wish the scattered, fractured breaths would cease, because we all know that they will.

And then they do. Though there have been several times when twenty, thirty, sixty seconds passed between desperate breaths, this time each of those segments pass, once, twice, and there are no more breaths forthcoming. I cannot tell if his parents notice. Then his father says, "He isn't breathing anymore, but . . ." and points to his son's abdomen, where the aorta is still pulsing, a sign that the heart is futilely pumping, refusing to cease. We expect it to cease momentarily, because, like any muscle, it will run out of oxygen, too.

I do not know how many minutes pass, but several. Perhaps four. Perhaps five. The father asks the nurse to close his son's eyes. She does. Another moment passes. And then, as if we were in a horrible movie, there is a long, loud gasp from Mr. L. I am so surprised that I jump. There is another. And another. And then they come more regularly. And he still looks lifeless and gray, but he is clearly, inexplicably alive. When, after ten or twelve minutes of breathing, it becomes obvious that Mr. L will breathe on his own again for at least a short while, the doctor and nurse explain this to the parents and then leave the room, and I follow them.

The nurse says in thirty years of caring for the dying, she has never seen anything like it. The doctor says one time: She left an elderly, demented man who she thought had just died after telling the nurse she was going to call his wife and have her come in. She says the man began to breathe again, and the moment his wife arrived and held his hand, he died for good.

Was Mr. L alive, dead, then alive? Was he dead until his breath resumed? What did I see him experience? I was not there when he took his final breath, but did I see him die?

During my first semester of medical school, I cannot know how the emotional difficulty of the actions we perform on our cadavers will help us prepare for the agonizing moments we will observe in the lives of the living. Nonetheless it is a welcome shift away from this intensity when the pelvic hemisection is complete and we move on to other dissections in lab.

The afternoon following the hemisection, we are to roll Eve over, dissect the deep muscles of the back and chip open vertebrae, with a chisel to look at the spinal cord. Now that Eve's left leg is detached, turning her body is significantly easier. While Tripler and Raj take the whole afternoon to snip at the vertebrae first with heavy-duty toothed shears called ronguers and then literally use a chisel to break into each whorled vertebral body, Tamara reviews cross sections, and I spend my whole day "cleaning up" the deep pelvis.

It is inevitable that when Dale or Dr. Goslow come over to answer a question, somewhere in their response will be "so just clean this up a little more. . . ." No matter whether we are looking to locate a structure, trace its path, find branches or innervation, or even understand its functionality, there is always more isolating to be done. There is rarely time built into the syllabus for this kind of detail, so when the assigned dissection is one that cannot be manned by everyone at the table, it is a good opportunity to tend to a previous dissection.

Clearing away fascia from structures takes patience, and the fascia clears away differently in different places. Its consistency—and hence its tenacity—varies. Under the skin it is webby and soft, and you can push through it with your flattened hand or a finger inserted between muscles. In areas of fat, it is also soft, but thick; using rat-tooth forceps, you can "pick" it away if it is in globules. If it is less dense, a whole area will fade into one clingy strand, much as a spiderweb does when you grab it with a broom.

Clearing this kind of fascia makes a sound like peeling a slightly green banana. It is quiet, and one of the parts of anatomy I find most satisfying. Progress is clear, and you don't have to battle; each movement is an act of revealing—the thick belly of a gray-brown muscle, the shiny white of tendon, the gloss of smooth muscle in an organ. You slide your hand beneath one of the four quadriceps and you can isolate it from its partners, hold it by itself, your hand wrapping under and circling all the way around it. You can slide your hand up to the point on the pelvis or femur where it begins and down to some slim tendon where it ends, and you can easily imagine that if the space between the two ends shortens, then the knee has no choice but to kick out, just as it does in reflex tests in the doctor's office.

Other areas of fascia are tedious. The thick aponeurosis of the palm is replicated on the sole of the foot, and it is very slow going in both spots. The connective tissue is tough and irregular, as one would expect from the functions of these all-important areas. In these regions I am impatient and use a scalpel to clear despite suggestions not to. I am sometimes left at the end of the day without a small nerve or artery, having "blown through it," as Dale would say. The texture of this fascia is tough and stringy, like the mesh coating of a scrubbing dish sponge invested with fat where the sponge would be, occasionally harboring vessels and nerves.

By the afternoon's end, I have separated most structures of importance inside the pelvis, having cleared (and thrown) away tangles of vein. We can see clearly how the common iliac artery branches into the internal and external arteries and gives off blood supply to the uterus, rectum, bladder, even through the pel-vis to the gluteus muscles and one muscle of the thigh. We can pull on the inferior gluteal artery and nerve and see them tug on the gluteus maximus muscle on the other side of the pelvis. More than anything it is exhilarating to me how much the afternoon and the search for branches among the mess helps me learn what is what and what is where.

I do not pay much attention to the spine dissection until Tripler and Raj are through. I can, after all, easily roll the leg around whichever way I want the pelvis to be oriented now that it is disconnected from the upper body on which they are work-ing. They are tired. An afternoon's work has given them just a three-inch window of the spinal cord, covered by a cloak of dura mater, a tough, membranous coating, translated literally as "tough mother," which protects the brain and the length of the spinal cord. The cord itself is unimpressive, given its extraordinary func-tion. Eve's is dark brown and about the diameter of my pinkie, dull in comparison to the sciatic nerve, which shoots out of the pelvis and down the leg and is the size of a sturdy rope and a gorgeous, glistening off-white. The colors of the spinal cords vary from ca-daver to cadaver, as they do with all tissues in all structures. A rather incredible look at a dorsal root ganglion has emerged, how-ever, complete with roots and rami, over which you can trace the path that sensory impulses would take to the central nervous sys-tem from the rest of the body. It is tiny and perfect and replicated on each side at each vertebra. Crazy.

At the end of this lab, Dr. Goslow holds a question-and-answer session, as the second exam is only days away. We all sit by our cadavers. I clean off little pieces of fat from the body and clear away obscured portions of muscle as I listen. This process that just weeks ago kept me from falling asleep at night is now something I do to occupy myself, strangely satisfying.

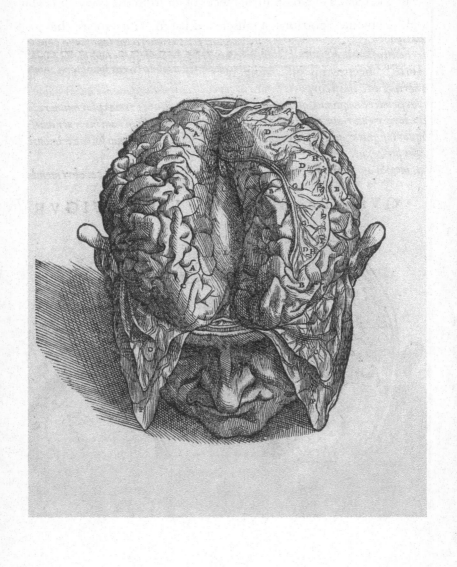

Dismantled

And did you get what
you wanted from this life, even so?
I did.
And what did you want?
To call myself beloved, to feel myself
beloved on the earth.

RAYMOND CARVER, "LATE FRAGMENT"

As the semester wears on into its final month, subtle changes in ability and perception occur in my classmates and me. Sometimes we notice them, such as when we dissect the leg and its structural similarity to the arm makes us laugh at how tentative our first cuts through the skin and fascia of the arm were, how befuddled we were by uncovering the musculature we expose so easily in the lower limb. Other times the change is a more whole and therefore less perceptible one, as when you are twenty and someone perfectly describes certain yellow orchid blossoms as miniature ball gowns, and it becomes impossible to remember ever looking at them without seeing the full skirt, the cinched waist. Eventually the thick elasticity of artery and the sheen of tendon are second nature to us. We cannot imagine not knowing that it is the pulse of the descending aorta that visibly beats in the thin woman's belly, the bony zygomatic arch that forms the cheekbone's rise.

And from the moment when anatomical knowledge begins to take on that quality of the innately known—at some unidentifiable point midway through the semester—the foundation is laid for an understanding of disease, and then a basis from which we can counteract the disease. In this way our dismantling of the body gives birth to our ability to make the sick and broken body whole.

Dr. Goslow takes advantage of this shift and tries to make us feel more settled in our new point of view. One day, in order to lighten things up in a lecture on gait, he shows us clips of Hollywood movies in order to discuss the topic. The movies are a perfect learning tool, liberating us all from memorizing isolated facts and reminding us that the big picture of anatomy is that the structures all actually work together. He turns down the lights and lowers the screen. A terrified woman in a knee-length skirt and matching suit jacket is running through city streets, looking back over her shoulder at someone unseen, chasing her. The next series of frames sends us into gales of laughter, as we learn that it is King Kong in pursuit, thumping along in his own earth-shaking run.

Dr. Goslow stops the tape, turns up the lights, and jumps up on a table in the front of the lecture hall.

"Okay, you guys, so we all know we run like this," and he makes the motions of running, leaning forward slightly but quite upright, pumping his arms and legs in a forward plane. "But Kong and his fellow primates, they've got an entirely different center of gravity." In a startlingly good impression of the primate body habitus, Goslow spreads his feet wide, toes turned out and knees bent, and hunches his shoulders. He ambles across the table, arms dangling, body low and squat. The room roars. Our distinguished professor makes a terrific gorilla. "Do you see the difference?" he asks. "Gorillas have a huge upper body, I mean HUGE, in comparison to us, so their weight is shifted, and therefore their gait is totally different than ours. All right."

The lights dim again, and Frances McDormand comes up on the screen in *Fargo*. She is rosy-cheeked in a bleak, snowy field with a dark blue police uniform and hat on, and she is easily eight months pregnant. As she trudges through the snow, she spreads her feet wide, and leans back, her hand on her lumbar spine. The clips continue: Jack Lemmon and Marilyn Monroe are both walking in heels in *Some Like It Hot*, leading to a discussion of gendered differences in weight distribution; a group of sailors belowdecks listening to the creepy thud of Ahab's wooden leg in *Moby-Dick* are used to demonstrate how the rhythm of gait is altered when there is no give in the ankle or knee. "Hunchback" Igor shows us the walk necessitated by a bent spine in *Young Frankenstein*; a young girl who suffered from polio wends her way with arm braces and outwardly curved legs in *The Little Girl Who Stole the Sun*. Finally, Kevin Spacey as "Verbal" Kint in *The Usual Suspects* leaves the prison office where he has been held and shifts from the palsied, stroke-induced gait of his alibi persona into a completely normal stride. The whole lesson is perfect, both in tone and in effectiveness. We have the images of these walks etched in our minds, via lightness and laughter.

In lab, as my classmates and I make our ways further and further through the remaining regions of our cadavers, the bodies begin to look less and less human. With fewer areas covered by skin, body cavities opened and emptied, and flayed muscles disrupting the familiar outlines of the limbs, it is sometimes difficult to tell from afar which body part each table is dissecting. It is during this stage that my parents come to town for a visit. I invite them to come into the anatomy lab if they like, to see the work that I have been doing, what a heart looks like, or a lung. My father endured the loss of his own mother at a young age. As a result, perhaps, death is particularly resonant for him and mortality a great fear. He chooses to come with us to the medical school but wants no part of the cadaver viewing. My mother, on the other hand, matter-of-fact and not at all squeamish, enters the lab with me and, unfazed, promptly states that the bodies look a lot like turkey carcasses.

When Raj, Tamara, Trip, and I have finished a rather quick dissection of the leg, only Eve's head and neck remain to open and explore. The head dissection gathers together many of the questions upon which the experience of gross anatomy has forced me to reflect. The brain is the true embodiment of my own conflicted response to anatomy. Somewhere deep within its crenellations, here lies wonder, and here lies the question of whether we have a right to pursue wonder in seemingly inhuman ways. Here is the knowledge gained by dissection, which drives our actions forward, and here lies the toll the process takes on each of us, in stress or dreams or dissonance. Here in the brain is the newly transformed identity of the doctor-to-be, with a beginner's knowledge of disease and healing, with a stomach more steeled to trauma and to death. But somewhere, too, there must be the echo of the person who existed before cutting a human body, before feeling the cool stiffness of a pulseless heart.

The brain—the mind—is the manifestation of the liminal spaces into which doctors plunge. It is where personhood resides, of ourselves and our loved ones, of silent Mr. C in the ICU and, though undeniably quiescent, of Eve. The brain is what we as doctors fight to save, whether we are nephrologists or orthopedic surgeons, psychiatrists or dermatologists. We are trying to heal and preserve personhood, which can reside nowhere else. Still, in the midst of all these truths, the answer to how I feel about myself now as a dissector is not eminently clear.

Perhaps in a shared acknowledgment that this fact is collectively true, the entire room quiets when the face dissections begin. We have all now removed the plastic bags tied around our cadavers' hands and feet, and though they are mostly replaced in order to keep the flayed structures in the fingers, toes, palms, and soles from drying, the mystery of what lies beneath these bags is totally gone. The bags over the heads, however, are a different matter. No group, including us, has spoken of their cadaver's face, so I have no way of knowing whether any of the other groups have done what Tripler and I did on the first day, whether anyone else untied the cord around the neck, lifted the head to pull away the plastic bag, unwrapped the gauze to look at the face.

Initially I take the quiet in the room to signify that no one has seen the face of his or her cadaver, that this odd unveiling is new to everyone but us. Then I realize that we, too, are quiet. This moment has no less power for us, despite the fact that we have seen Eve's face once before. Throughout the room lengths of gauze are being unwound and shorn heads and faces surfacing. What strangeness. The mix of reactions is inexplicable. How are you supposed to feel the moment you first see the face of a person whose body you have cut into, cut apart? A body whose interior you know more completely than your own or that of any loved one? Certainly we feel different about Eve now than we did that first night in the darkened lab. Her beautiful dead face is the same—mouth slightly open, eyes slightly closed. Only now her face seems to float above a dismantled body. The context is altered. She has become more and more like some dark, strange work of visionary art: opened chest, emptied belly, leg cut through and rolled to the side. And she is *Eve* now; large-stomached, gallbladderless, slight-framed Eve. It is strange to think that I feel I have come to *know* this woman. I've seen her daily, touched her, cared for her, been pushed and angered by her, dreamed about her, but *know* her? I know her body, yet in a manner that a psychoanalyst would love: Any emotion, any perceived relationship, any tenderness or fear or repulsion or frustration, any *humanness* is my own and only my own. She knows me from no one. She knows nothing. Anything beyond the physical is my own construction. I do not even know her name.

And yet when I look at her face, I think, *Yes, that's Eve, but now less so than before*. Because if I have been witness to anything in her life, it has been this final change, where even the physical evidence of her life, left long ago by any spirit, now resembles her less and less. And it is hard to know just how big or small a role our dissection plays in that process.

One November I flew west to be with a dear friend at the funeral of her twenty-four-year-old brother who had hanged himself in their parents' basement. Even in the casket, days after he had died, he had the beauty everything young does, and it was that beauty that struck the chord of most powerful dissonance. Amid lilies and vaulted ceilings and song, here was his lovely face that dared anyone to say, *Better off* or *It was his time*. But even with that beauty, even with his hands that were identical to my friend's, even then he resembled himself less and less. From the calling where I said—and meant—my first Hail Marys to the moment just before the casket closed when I was seated in the church and from behind me heard my friend wail as if she had just found him hanging, already he was further gone.

I think of the Jewish sense of what happens after death, that we live on only inasmuch as we are kept in the memories of the living. And if we have the chance, the totality of our memories of people includes how they are with us after they are dead. My friend's brother was less himself over the course of our days of mourning because our sense of his living self as something with us lessened.

He is, of course, kept well in our memories, but kept differently now. The fact of his death intercedes. I never remember him as the living person he was. He is locked at twenty-four or younger, tainted by tragedy, past rather than promise.

And this is not unlike the shift that has occurred for me the second time I see Eve's face. I recognize her, know it to be her, but know, too, that the changes I have witnessed—*caused*—in her are irreversible ones. That she will be kept differently in my memory. And as my friend cried out at the moment when the casket was just about to close, I understand that even these moments of reference are tenuous. In the next days, I will peel away the skin of Eve's face. I will open her skull. I will render her unrecognizable. More than any other part of her body I have altered, her face, as it is for all of us, is her identity, and I will remove it. This—the first chance for many to look at the faces of their cadavers—is the last chance for us to look at the faces of our cadavers.

I wonder, then, what this brief moment feels like for those who have not looked beneath the gauze until now. Are they surprised? Does the face seem to fit? Does the group with the muscle-bound, tattooed cadaver think his face is smaller and sweeter than they would have thought? Does the face of the woman with lavender fingernails have jagged teeth—or no teeth? I suspect there are those who look at the face only as quickly as they can, or not at all, other than to perform the dissection. Though no one admits to it, I suspect there are those for whom the face is too real to focus on, too human.

But for most of us, when we have finished looking at the faces of our own cadavers, there is a kind of furtive glancing to see what other faces look like. We think there will be a power in this—that we will be moved by the emerging individuality of the cadavers—but the truth is, there is not. Partially this is due to the fact that, like the breathless bodies when we first saw them in their bags, the faces of these people do not look like faces of living people. The color is wrong. Their eyes are half shut, or one eye is closed more than the other. As was true with the first cadaver I ever saw, some of the faces have flattened cheeks or noses—postmortem alterations that make it impossible to picture what their faces were like in life. Their faces are, in shape and color and expression, like no living person's face. Partially, too, the lack of resonance in the viewing of the faces is the result of the androgynous appearance that is unpredictably but universally lent to the cadavers by their shorn heads.

During the below-the-neck dissections, when we would study cadavers other than our own, we consistently used gendered pronouns in our conversation. "Does he have a good spinal accessory nerve?" or "Her subclavicular takes this crazy turn." In what proves to be one of the more paradoxical realities of the semester, people increasingly misidentify the gender of other cadavers after the heads are uncovered. This is perhaps less surprising once facial dissections are performed, but even when the faces are intact, the shaved heads and un-made-up faces are, for the most part, gender neutral. Even Eve, so feminine in the fineness of her features, is frequently referred to as "he" by approaching classmates. "Can I see his digastric muscle?" they say, and we cringe.

"Her," we say.

"Sorry," they reply. "Can I see it?" Inevitably, we make the same mistake at other tables on other bodies. Without even realizing it, some of my classmates begin referring to individual bodies in the genderless, if plural, "they." As in, "Do they still have an intact phrenic nerve?" Something lessens.

The night before the head and neck dissections begin, another dream: I am at my desk, holding Eve's brain in my hands. I have two or three atlases spread out before me and am locating various divisions and compartments of the brain. *Caudate. Putamen. Corpus callosum. Septum pellucidum. Fornix. Anterior and posterior horns of the lateral ventricles.* I race through them over and over again, growing increasingly distressed. *Lateral sulcus. Genu of the corpus callosum. Lentiform nucleus. Infundibular recess of the third ventricle.* I am missing something. I frantically turn pages, look at diagrams, read charts, look again at the brain.

I cannot find her memories.

At this stage in the semester, our *Dissectors* have become disgusting. Each time Tripler has to consult the book for its instructions, she grimaces. "Most foul," she groans. Days on end of gloved hands covered in formalin and the grease of fat, with the odd clinging particle of dried brown blood or unidentifiable tissue, have flipped through pages and rendered them nastily translucent and endlessly damp. The *Dissector*'s instructions sounds as if they were for something as banal as a cobbler recipe. "We will begin the dissection of the head and neck region by exploring the muscular and visceral components of the neck first. Then we will proceed onto the face, remove the brain, study the orbital region, and then work our way systematically down through the head toward the deep structures of the neck."

"Oh, nooooooooo," Trip wails, and I share her sentiment. We know better than to trust the facile tone of the text. "As if we can just—poof!—remove the brain," she says. "And by 'study the orbital region' I doubt they mean just lean over and look closely." We both give a kind of defeated laugh, knowing that she is right and that it doesn't matter. That we will open Eve's skull and remove her brain, dissect her eye and socket, and "work our way systematically down through the head toward the deep structures of the neck," whatever surely ominous path that implies.

We make a series of cuts, starting with the scalpel behind Eve's right ear, guiding it down to her jawbone and along the edge of her jawline to her chin, then continuing, to make the same cut on the other side, finishing behind her left ear. The bone is just beneath the surface, so the splitting of the skin is visible only if you look closely. The skin now has a kind of reverse seam, two new edges. The other cut is perpendicular, beginning at the chin and sliding down the center of Eve's throat, ending where, on the first day, we peeled away the skin across her clavicles. As we did with the skin on her chest, we grasp the corners of skin we have made at her chin (how inherently unlike skin to have edges, corners) with the rat-tooth forceps. Using the scalpel, we cut away the connective tissue beneath, little by little. The skin peels back, revealing the crowding of muscles in the human neck.

Because the neck lacks the bony protection conferred by a rib cage or a pelvis, for example, many of its structures can be seen or felt on our own, living bodies. The most easily identifiable muscle in the neck is the sternocleidomastoid. The name sounds prehistoric and powerful, but in reality it merely describes the bones connected by the muscle's fibers. The sternocleidomastoid originates on the mastoid process of the skull, the bony prominence you can feel immediately behind your ears, and attaches on both the sternum and the clavicle. The muscle is clearly visible on a slender-necked person and can be quite prominent when the head is turned. In order to feel the tendon, we each return to our own jugular notch, which was our original landmark on Eve before we made the first cut across her sternum. We raise our fingers again to the very bases of our throats and find the notch in the bone. If, with your index finger in the notch, you then turn your head so that you are looking over your right shoulder, you will feel that your finger is now held between the bony notch and the tendon of your left sternocleidomastoid. You can grab onto the tendon and follow its course up the side of the neck, feeling the belly of the muscle. The muscle becomes even more prominent when flexed. To feel this, again look over your right shoulder, fingers along the sternocleidomastoid, and then, keeping your neck twisted to look over the right shoulder, tilt your head so that your left ear is parallel with the ground. The tendon goes taut, and the muscle is easily palpable.

Because the face and head dissections are the final ones of the class, I think we are all surprised by the extent of the reactions we have to them. Lex, heretofore a stalwart mix of studiousness and piercingly smart humor, leaves the room in a rush. He explains to me later in a quiet tone that he had a panic attack; he felt his heart would pound out of his chest; he felt he could not breathe.

Despite the inertia of the dead, they actively affect us. Even now, as the term winds to a close.

There is, of course, an opposing argument. Lex's panic does not stem from his cadaver; she has, of course, done nothing differently today than every other day he has come to her side. Rather the fear comes from the fact that he is in a room full of otherwise relatively normal people, his friends, his colleagues, and we are all engaged in taking the faces off dead human beings. Some cut through lips with scalpels. Others pull off masks of skin so they are holding in their hands the obvious oval of nostrils, whiskered cheeks, eyebrows. When they are not cutting, they talk to one another, consult their textbooks, and as they do so, their hands rest on a dead man's chin, their forceps sit across his forehead. And how much easier for me to write of the things "they" do. As if my thumb does not come to rest on Eve's eye socket or lips or teeth when I am trying to identify the muscles with which she used to chew.

The upsetting nature of these dissections resonates even more deeply with me because they occur as my grandmother recovers from her stroke and as my grandfather begins to die. My grandmother has filled a space in my life that no one else possibly could have. She is, above all else, a truth teller. She is the kind of woman who cannot see the point of beating around the bush, who does not understand how talking around something could possibly be any more effective than saying things the way they are. She is rare in this way as a person and even more rare as a woman, and there is something deep inside me that responds to her whenever I am near her and says, "Yes, yes."

Unlike me, she is not at all sentimental. Deborah loves the story of when I asked her for the meaning behind my names. My first and middle names were names she had chosen as middle names for my mother's two sisters, so I was anxious to hear from her how she had chosen their names, which then became mine.

"Well," she said, "I always wanted to name a daughter Christine, but when the neighbors at Higgins Lake beat me to it, I certainly didn't want to do the same thing they had done, so I just kept it as a middle name. As for Elaine, I guess by the time I threw out all the names of people I didn't like, that was pretty much what was left over." Though it wasn't the emotion-laden story I might have hoped for, I knew it was honest, and so, crestfallen as I was at that moment, I have come to love the story, too. It represents her perfectly.

She was the college freshman at Michigan who danced with an older sorority sister's date at the pledge formal and hence landed my all-American basketball player, law student, hotshot catch of a grandfather. (He would always tell the story of how he came to pick her up one Saturday evening for one of their first dates, only to find her coming home from an afternoon picnic with another beau. She, unapologetic, said good-bye to the afternoon date, ran upstairs to her room to brush her hair, and came back down, ready for an evening with my grandfather.)

In my childhood she was the grandmother to whom you brought your bluegills when you caught them in her pond. She would take out the hook for you, whack them on the head with a stone, fillet them in the laundry room, and then cook them for that night's dinner alongside fresh Indiana sweet corn and garden asparagus. So it was through her that I first learned about death—about the birds that got tangled in the nets around the blueberry bushes, about the snakes that got hacked in half with her spade, eventually about the emotional fatigue of aging: "It's a damn shame when every single one of your best friends is dead," she would sigh. And if she said it was, I knew it must be true, though it was—and remains—nothing I could imagine.

During World War II, Grandpa was playing semipro basketball and practicing law, and so he got sent to Texas to be in the Judge Advocate General program instead of to more dangerous places. That left Grandma in Indianapolis, a woman in her early twenties, living in the house next to her in-laws, with four children under the age of four. Never a woman to want too much input from someone else, Grandma tired quickly of this arrangement and packed herself and all four kids into the car, driving the eight-plus-hour drive to Michigan, to Higgins Lake. Once there, she lived with the children in a cabin with no heat or electricity or running water. She heated bathwater over a kerosene flame and washed diapers out in the lake, and she preferred the situation mightily to the one she had left.

"What a dumb thing to do," she would say of her time alone at the lake with the kids, looking back years later. Of her mother-in-law, she would claim, "I'm sure she said one little thing I took offense to, and it got my goat, and without giving it any thought,

I just decided that was that," but her smile as she said it always made me understand that she knew exactly why she had done what she had done and would do it all over again the same way. The four children, the packed car, the lake, broad and dark when the sun went down, crickets and tree frogs singing through the screen door, through the starry clarity of Michigan nighttime.

A grandmotherly figure she was not, but a figure bigger than the room, a figure you couldn't help but fall in love with, couldn't help but want to *be*.

The story of her life with my grandfather is, above all, a love story. From the pledge formal to their dating days in Ann Arbor, to their raising four children together and eventually growing old together, their devotion to each other was plain, as was their steadfast belief that they had each scored big in landing the person everyone else wanted to catch. Photos of my grandparents at various stages of life are all ravishing and depict a life that ranges seamlessly from daily midwesterndom to worldly glamour. At a bridal shower for my mother, my grandma is stunning in a yellow sixties minidress and white hoop earrings. My parents' wedding was at my grandparents' Mooresville, Indiana, home, and my grandfather looks regal in his tuxedo and perfectly combed hair, welcoming guests onto the porch for cocktails before dinner, to look at the sculpted ice swan and listen to music from a brass band. In another shot he and my grandmother are beaming and tanned on a deep-sea fishing boat in the Caribbean, holding a huge, shimmering tuna between them. At their fiftieth anniversary party on my uncle's porch, my grandfather diligently looks at the camera and smiles, while beside him my grandmother nearly cackles with laughter, looking just past the photographer at something no one can now remember.

In my favorite series from 1941, my nineteen-year-old grandmother and twenty-three-year-old grandfather are at Higgins Lake, he in a dark pin-striped suit and diagonally striped tie and she in a polka-dotted shirtwaist dress with a knee-length pleated skirt and fabulous white-and-black spectator pumps. World War II has begun, and they've married quickly, before my grandfather has to leave for JAG duty. The family lake cottage will be their honeymoon retreat. They are seated in one picture and holding hands, both gorgeous, with the openmouthed grins of newlywed disbelief. My grandmother looks as if she will break into a gale of laughter at any moment. In another shot they mock solemnity, my grandfather seated with his hand over his heart and my grandma standing beside him, one hand on his shoulder, one gripping a clump of wildflowers. Neither of them smiles. In a third, a commentary on the speed of the wedding, my grandmother's parents have orchestrated a hilarious scene. My young great-grandfather holds a shotgun and looks purposefully at the couple. My grandmother has a towel tied to the back of her head with a belt and holds a bouquet of cattails. Beside her my grandfather takes a swig from an old whiskey bottle. My great-grandmother, barefoot, buries her face in a dish towel, weeping. A family friend holds open a dog-eared Bible and wears a neck brace meant to resemble a religious collar. Each of their expressions is pure seriousness. I can only imagine how many shots were ruined by laughter. Knowing what I know, it is nearly impossible to look at the faces of my grandparents in these pictures for what they are—near children, in a moment of pure joy, unaware of the expansive family they will begin raising a year from then, of the sixty-year relationship into which they have just entered.

It has been clear, always, as one of those things that every family member knows but may never discuss, that my grandmother might survive many years should my grandfather predecease her but that the inverse could not possibly be true. In the years before I enter medical school, my grandfather has increasingly faced the battles of aging: breaking hips, losing his breath to worsening congestive heart failure, harboring the cool, weak-pulsed feet that are harbingers of failing circulation in his legs. His health has been diminishing in seemingly unrelated ways that add up, even knowing what I know now, to no diagnosis other than advancing age.

Once, having raced to Higgins Lake after my grandfather fell off the metal dock and into the shallow water in an attempt to step into a boat to go fishing, I sat with my grandmother in his hospital room, waiting for Grandpa to be rolled off to hip-replacement surgery. When the surgical transport team came, unlocked the brake on his bed, and began to wheel him away, Grandpa looked tearfully up at my grandmother and said, "Didn't we always have fun with our Sanka after dinner?" He would survive the surgery and, despite his abysmally low motivation for rehabilitation of any kind, would survive many years beyond it. But that moment remained with me, a vision of the patriarch afraid, who, when pressed to define what was meaningful to him over the course of his life, looked to the woman he had loved so long and spoke of their nightly routine of ten or fifteen minutes of decaf and conversation. The wisdom of that moment of pure emotion resonates with me still: that the dearest and most enduring moments of our lives are sometimes the quietest ones.

The summer before I began medical school, just before my grandmother had her stroke, my grandfather awoke in the middle of the night with a cold and painful leg. He displayed the symptoms that I would later learn clearly indicate femoral arterial occlusion. A hospital evaluation did not take long to determine that my grandfather needed surgical intervention. The pain was caused by early necrosis. A lack of blood flow to the tissue meant that beyond the blockage the leg was receiving no oxygen. Without oxygen the tissue was beginning to die.

My grandfather underwent surgical bypass, grafting a vein to his femoral artery above the site of the arterial blockage and reconnecting it below, thereby providing an alternative route for oxygenated blood to reach his lower extremity. The surgical wound, by necessity from harvesting the vein and accessing the artery for the bypass procedure, was massive, stretching from his groin to below his knee. Thankfully, his fragile body tolerated the stress of surgery and anesthesia, but, to my grandfather's chagrin, he was sent to a nursing home following the surgery for rehab therapy and skilled nursing before he could be sent home.

When, therefore, his wife of nearly sixty years had a stroke and was hospitalized, my grandfather was in his own state of recovery. And because, in her rare moments of awakening, my much-changed grandmother asked clearly and repeatedly to see her husband, my grandfather received special permission from his geriatrician to be driven to the hospital and wheeled in a chair to my grandmother's bedside.

My mother and I had grave concerns about how my grandfather would handle the sight of my previously strong and out-loud grandma in her current condition, bedbound and only intermittently lucid. My first sight of her after her stroke had racked me, and I had had to take hold of the hospital bed frame to smile at her and remain standing.

A journal entry from those first days:

She is able to speak and is totally lucid, with the exception of moments in between sleep and wake. The second or third day, she told me a little boy had found pieces of her eye, and now they were at the fire department. A few days ago, she woke up thinking she owed money to a bunch of people on an airplane. Yesterday she at least knew it was a dream—but stayed stuck in it awhile, looking alarmed—and had been dreaming of "spare parts"—I think prostheses—that "didn't fit" and were "like lobster claws." Shortly after that, though, she regained some motion in her left leg (though continued to bemoan the "left arm not worth a damn"). What a week. It was simultaneously ugly and wrenching, but with these moments of small progress that were somehow now relatively wonderful and important. The first few days were very tough. I arrived to see her with oxygen in her nose and eventually a feeding tube in her nose also, which was excruciating to put in and which she pulled out in her sleep one night. Then she had an abdominal tube put in, through which she was fed a Slimfast-looking bag of stuff laced with a tincture of opium to try to stop her diarrhea. I had to adjust to an extremely weakened Grandma, who spoke rarely and softly, slept all but five or ten minutes an hour, opened her eyes only barely and only if asked to, and flopped around lifelessly as nurses rolled her, pulled her on and off her bed and gurneys, etc. The first few days, she would be so exhausted that she would sleep—even snore—through nurses turning her body, starting IVs, giving her shots in her belly, pulling her onto a gurney, wheeling her on and off of elevators, and she wasn't even medicated heavily.

We first tried to gently suggest to my grandmother that she wait to regain a bit of her strength before she had a visit from my grandfather. Wouldn't she rather wait until she could be assured of staying awake while he was there? But she asked for little during those days, and her request for him was consistent. Eventually we realized we could not guarantee that she would be any better a vision for him to see days, or even weeks, from then. Even more important, we knew that she had undergone this substantial loss, was afraid, was having terrible dreams that disoriented her, had no idea of what the future might hold, and wanted her husband. It was such an obvious desire.

My mother, who had kindly tried to shield Grandpa from some of the severity of Grandma's condition, fearing that it would slide him in to a deep depression that could be a blow he might not recover from, began to prepare him for what he would see when he saw his wife. She also tried to explain to him that Grandma's state was tenuous and that it would be important for her that he not have too severe a reaction, which could discourage or upset her. Grandpa, consistent with his nature surrounding difficult conversations, was mostly quiet on the subject. In her characteristic thoroughness under such circumstances, my mother consulted with a social worker friend who specialized in geriatric issues. The advice she received was simple but good. Keep the visit relatively short, and then take your father to ice cream afterward. Keep conversation light, but try to get him to talk a little bit about his feelings after having seen her. We had no idea what to expect, but it would be dishonest to say that we did not fear a mutual collapse.

As it turned out, I could never have predicted what actually transpired. I left Grandma's room to meet my mother and Grandpa when they arrived. He appeared solemn and nervous, his postoperative leg heavily bandaged and his tall, athletic frame crunched into a wheelchair. We rode in the elevator up to Grandma's floor, and my mother and I made small talk in the elevator about the tiny but encouraging bits of physical progress she had shown that morning. A bit of movement in the left leg. Slightly longer periods of wakefulness. Grandpa said nothing and looked straight ahead at the elevator doors. When we reached the ward, I walked quickly ahead to tell Grandma that they had arrived and to make sure she was awake and as presentable as possible. She was a little tearful.

Suddenly we heard a booming voice outside the door, coming down the corridor. "Is this the room of Mary Townsend? Magnificent wife of John Townsend? I'm looking for Mary. Is this where I can find my love?" We looked to the doorway, and in came

Grandpa, wheeled in by Mom and gesticulating dramatically. In a moment for which I will always love him, at first sight of her he registered nothing like sadness or disappointment or fear. Instead he looked squarely at her and said, "There is my Mary. Mary, you look so beautiful! I've never seen you more beautiful in all my life." I, on the other hand, had never seen such an outpouring of emotion from my normally reserved grandfather, and he was nowhere near done. Both of them in tears, he took her hand in his and began to talk to her about the picnics they used to have in the arboretum in Ann Arbor, where they would lie on a blanket and kiss. He asked my mother and me to help him stand, so he could lean over the rail of the hospital bed and give his wife a kiss. He began to sing, in a full voice, a song to her about the wind in her hair. In so many ways, the scene was implausible: My grandmother was, at that time, a fragment of her former self, exhausted, sick, with plastic tubing in her nose and side and IVs in her arms, and my grandfather was as vociferous as I had ever heard him about her beauty and allure. It was, without question, the most perfect thing he could ever have done for the woman he loved, and it will remain, I suspect, the image that defines the depth of true love for me for the duration of my life.

Over the course of the fall, my grandmother's health slowly improved and my grandfather's health gradually declined. I was aware that my own personal grappling with the body's ability to heal and to fail colored the ways in which I approached my medical studies. At school during the day, the minutiae of histological slides and physiological equations seemed too far removed from the medical treatment of real people. At night, during phone conversations with my parents or grandparents, I felt the full thrust of how illness affects people's lives, but I was not yet equipped with any of the knowledge to help assuage it. The result felt cruel and misdirected, like learning the alphabet and numbers in a foreign language, when you just want to know how to ask for the bathroom.

Rationally, of course, I knew that the first year of medical school was a necessary foundation for clinical knowledge. And similarly I knew that even as I would acquire a broader and deeper understanding of medical treatments in the years to come, there was nothing that I could do differently from what my grandparents' fully trained doctors were doing. None of us knew the cure for aging. None of us could prescribe medication to return an elderly couple to the sepia-toned photograph in my living room of barely twenty-year-old newlyweds, holding hands in a lake cottage and laughing at whatever the future might throw their way.

Still, the anatomy lab was a tangible space in which the specter of my loved ones' mortality swirled. For all the time spent wondering what was going on in their sick bodies, I could look into Eve and attempt to localize their bodily failings. The dissection of the head would have already held this potency for me, due to the fact that a clot of one of the brain's vessels had taken the use of my grandmother's left side, along with an unlikely collection of numerically oriented skills. She could no longer tell time. Even looking at a digital clock, she did not always know the time of day. She could not play cards. She called my mother endlessly, complaining of the channel numbers on her TV being changed, of her VCR remote control not working. I wanted to see the vessel, named by her doctors and identified on CT scans and MRIs, whose blood supply momentarily ceased, blackening a section of her brain.

Less easy to locate, less easy even to attribute causality to, was my grandfather's physical and mental deterioration during this time. I could follow on Eve's leg the exact vessel in the leg that had failed him and the vein that had been used to bypass its path. I knew the track of the wound along his leg, and, when it failed to heal, I knew the line the orthopedic surgeon must have used for the amputation. But in what locatable space was the reason the tissue did not heal? I could find the vessel that occluded and brought my grandmother into this shifted and foreign existence, but which section of the brain explained these new moments of delusional fear in my previously steady grandfather when he thinks that my grandmother is having an affair with his nurse, when he thinks that the IRS has taken his remaining leg?

I am thinking about the brain, the look and feel of it. I am thinking about how the heart was at first mysterious to me, but then I could name its spaces, chart the flow of blood through it, fill it with water and watch its valves expand and catch. I learned the physiological formulas that described the electrical impulses in cardiac muscle, which could account for its incessant beating, for the life-sustaining propulsion of blood. I think of the lung and its branching network of exchange, oxygen for carbon dioxide, new for old; the kidney and its countercurrent exchange, serpentiginous nephrons, giving back to the body what it needs to retain and taking from the body what it needs to expel. The spleen, the pancreas, the stomach, the liver, the eye. All of it, in the past few months has been diagrammed and parceled, explained and organized. Demystified.

And now, though I can map the regions of the brain, draw its ventricles, trace the path of cerebrospinal fluid, even fathom synapse and neurotransmitter and nerve impulse, I cannot help but feel as if, when it comes to the recesses of the mind, I know a laughably small amount. My grandfather's mind is letting go. And that is the saddest trick of it.

What anatomy can explain to me the flight of dementia? What nomenclature can be given to the mean truth of a dying man couched in disorientation and fear? When I look at eighteen brains held in the hands of my classmates, I cannot differentiate one from another—not even in the way that one heart varied from another, or muscles did, or bones. Where, then, in the crenellations of the brain's tissue is the explanation for how a man's reason can depart when half of his leg does? Where is the space in which his firm knowledge of his wife's pure devotion gets lost? Where are the origins of the nightmarish visions that came to my grandfather in the interspace between life and death?

This—this seeking, based in both resolve and uncertainty, is unlikely to have any effect whatsoever on my interloping emotions from the outside world or my ability to cope with them. And yet this seeking is what I bring with me to the anatomy lab when we open Eve's skull and remove her brain.

On the day when we are to open the skull, Trip ties a string around Eve's head in order to draw a markered line around its circumference to guide the bone saw. We talk about the awful smell as the saw shaves a line of skull to dust and how the odor of bone is far more overpowering than it was in either the rib or the pelvic dissections. Tiny particles of bone float all around us, and we have the odd knowledge that we are inhaling specks of bone from Eve, indeed from all of the cadavers in the room, from whose heads rise wisps of smoky dust that dissipate until they vanish.

Moments like these, which are sometimes too much to dwell on, are perfect opportunities for philosophy, scientific theory, even poetry. If the whole of the universe was created from infinitely compressed matter in the big bang and if, as scientific law holds, matter is made only from existing matter and no truly new matter is ever created, then there should be no discomfort whatsoever with the knowledge that motes of another human's body are entering my own. If I can, happily and without hesitation, accept that the molecules that now reside in my own ring finger, hip, and spleen may have traveled first from a star and then billions of times over between sea and rain cloud and frog gut and branch and soil and lambswool and ocean floor and glacial shale or any such imaginable combination, then I shouldn't shudder to think that this palpable liberation from dead body to living one occurs at this moment, just as at any other. Three years, almost to the day, from the initial skull dissections, I will stand at the border of Higgins Lake, hoping to be pregnant, breathing in the damp

November air. Though by this point I know more than enough embryology to know that even if some hope of life grows in me, it is only a few cells and nothing more, I breathe deeply. I tell the hope to take the oxygen I give it now and hold tightly to it—that the place it comes from is holy and so is the perfect sustenance for these few cells that will become the foundation of all others. As if we, at our small and singular moments in time, can discern the provenance of the air we breathe and hence judge it more or less heartening, more or less wonderful.

Bits of philosophy like these are what Trip and I talk about as long as we can, in an attempt to will away the smell of the bone dust. When we are done sawing and must take a hammer and a chisel to the line we have made to begin to remove the crown of the skull, Trip comes undone. What I did not know until that moment was that a few years before she and I began medical school, Tripler's best friend left her apartment one morning to get coffee and a newspaper and was struck by a car. She was thrown and hit her head on the curb, which caused intense cranial swelling and left her brain dead. Trip saw her friend a last time in the hospital, with her head greatly swollen.

It is an agonizing conversation to have, particularly under such exceptional circumstances. We take an early lunch, which we eat indoors due to the inclement fall weather, but, desperate for perspective and fresh air, we bundle up and take a walk down to the corner coffee shop, where they make terrific chocolate chip cookies. We buy one to share and walk aimlessly, just to be outside and walking. We half laugh about how pretty the quad is and how we never notice, since we are only ever in the contemporary brick science building on the periphery of the undergraduate campus. One of the buildings we walk by has a row of large tinted windows at street level. In it I see our reflection—our ratty sneakers we have resurrected from dark closet corners so as not to ruin any worthwhile shoes in the anatomy lab, our thin hospital scrub pants beneath our winter coats and folded arms. I point it out to Tripler to show her how ridiculous we look, and for a moment, when we both face the mirrored glass, all I see is two young women who look deeply, deeply tired.

When we return to the lab, we have more chiseling to do, with the crown of Eve's skull still incompletely removed. The only word we can think of to describe our actions is "barbaric." Dale comes over while I am striking the chisel and says, "Ah, bone is her medium!" Neither Trip nor I laugh. Usually his jokes are funny and welcome, but today they strike us as irreverent and flip. We have run the gamut of emotion already today and have arrived at irritability. Our unease is compounded by our solitude. Tamara is home sick. Raj, after inexplicably missing his fourth lab the day before, leaves just prior to the skull sawing, and we are unforgiving. It is a task that has demanded physical and emotional divvying-up.

In fact, what we do not know then is that Raj will fail two of our three courses and will not even take our final exams. In hushed tones his friends will allude to the fact that he is suffering from depression. Distracted as I have been by his bravado during dissections, I will be stunned at the semester's end when I learn that he is one of a small handful of our classmates to drop out of medical school.

When we finally have chiseled through all the remaining bony connections between the crown and the base of the skull, we twist the chisel within the crack in various positions, in order to broaden the opening and try to remove the bowl-shaped top of the skull. We do this carefully, not wanting the chisel to slip and punch through the slitted opening into the brain. With each turn of the wrist, a kind of groan issues forth from the separating bone. The slit broadens, but the skull still holds fast, so we start to use the chisels to pry the top from the bottom. We use more force than it feels like we should need to, and Trip and I both grimace quietly. Finally, with a crunch, the top is freed, and now, in order to see the brain, we must only cut through the dura mater, the outermost protective layer of the brain, which attaches to the inner lamina of the skull.

The translucent dura mater feels a little like a thin layer of nylon, flexible yet resilient. It cloaks the brain and extends through the foramen magnum at the base of the skull to become the tubular sheath that encircles and protects the spinal cord. It also dives between the brain's hemispheres, creating a fold known as the falx cerebri. Before we can remove the brain, the falx must be cut away from the skull, and hence the dramatic midline division it provides cannot be appreciated. In order to show us the fibrous partition, Dale will show us a prosection that comes to be known as Falx Man.

Just after we remove the domed skullcap, we are called into the prosection room. Typically, the prosections are laid out on the same stainless-steel dissecting tables on which our cadavers lie. Sometimes they are entire bodies with several areas expertly dissected. Other prosections are isolated parts: an arm and shoulder, a single leg, a limbless torso. When we enter the room on this day, no such specimen is visible. We collect in a circle in the middle of the room and wait for Dale. A moment later he enters, empty-handed.

"Hey, guys," he says warmly. "How go the heads?" Mostly we shrug in response. Trip is more candid.

"Ugh. It's dreadful, Dale!"

Dale smiles. "This is definitely the rough stage," he replies, "but it's about to get really cool." With that he is ready to go. "Okay, so this is the falx prosection," he begins, and reaches down toward his feet, where there is a lidded white plastic bucket.

"Oh, noooooo," Trip groans. Her overt reaction is the one all of us are having internally. We know that the combination of head dissection and a prosection that fits in a bucket is an ominous sign. As feared, Dale quickly snaps on latex gloves, reaches into the formalin contained in the bucket, and lifts out a man's head.

In order to show the midline falx, the two upper sides of the skull have been sawed off and the hemispheres of brain removed. A two-inch strip of bone is left behind, running from between the eyes to the back of the neck. The man's face is totally intact. Dale holds the bony strip with one hand, making the odd impression of a handled basket. With the other hand, he points to the falx, which runs down from the handle and then, forming the portion of dura known as the tentorium cerebelli, spreads perpendicularly to either side, where the base of the brain's hemispheres would have been.

Various other structures are visible from this view, and Dale is pointing them all out to us when a classmate of mine leaves the room to be sick. It is the first time this kind of reaction has been publicly visible, and when she returns to lab the next day, she is embarrassed and explains that she had been fasting for Ramadan, and otherwise she is sure such a thing would never have occurred.

"*I'll be fine. I'm tough.*"

"*So yeh are, then, luv. So yeh are.*"

Those who have seen the brain in surgery say that the consistency of the live brain is somewhere between mucus and Jell-O. That it is gray. When we go to remove Eve's brain, it is firm and cream-colored. To detach the brain from the body requires us to cut its many connections, beginning with the spinal cord at the very base of the skull and including many critical arteries and nerves. Severing such important structures feels counterintuitive, and we consult the *Dissector* over and over to be sure that we have understood it correctly.

Reassured by the instructions, I slip my hands beneath the brain and try to lift it while Tripler gingerly aims the scalpel toward the connections we need to cut. The space in the skull is cramped and dark, and we proceed slowly to be sure that Trip does not cut something important, including my fingers. In order to give her enough room and as much visibility as possible, I am pulling the brain toward me. With each nerve and vessel cut, the brain inches steadily away from the bony base of skull, yet even once all the necessary attachments are severed, it does not emerge easily from its carapace. The *Dissector* instructs us to "cut . . . each of the remaining pairs of cranial nerves before gently 'delivering' the brain from the cranial cavity." Having as yet neither reached our months of hospital obstetrics nor birthed children of our own, Tripler and I are surprised by the degree of tugging required for the "delivery." Eventually, and not without some disconcerting tearing sounds, the brain is freed intact, and we hold it in our hands.

The experience is surreal. The brain, with its ridges and valleys called sulci and gyri looks like a contradiction: half prehistoric, half incredibly complex. Veins and arteries trace over its surface, contained in the space between the pia mater—"tender mother," the thin, transparent membrane closest to the surface of the brain, barely perceptible, clinging to the brain tissue as it does through all of its dips and curves—and the arachnoid, whose extensions up through the dura gave this meningeal membrane its spidery name.

Removing the brain has exposed the stumps of the numerous vessels and nerves that Tripler cut, poking up through their particular bony pathways in the base of the skull. In addition to learning the identity and function of each, we have discovered a new geography of the skull that has surfaced, with foramina and fossae and protuberances galore. Too fatigued to acknowledge that new realm, we allow ourselves to be satisfied with our freed brain and call it an early day.

When I get home, I take a scalding shower. I scrub my hair, brush my teeth twice, inhale water in my nose until I choke to try to rid it of the smell of the bone dust. That night, at home, Trip calls me to check in. She says that after lab she sat in her car and cried. When I hang up the phone with her, I open the window, despite the cold, misty night, and take deep breaths through my nose. I cannot get rid of the smell. I feel ashamed at being disgusted at this moment by the bodies, their skull dust, their skinless eyes. I feel ashamed, because I understand the unthinkable gift I have been given and how it deserves to be met with steady appreciation and reverence. I am ashamed to feel disgusted. But I am.

My journal from the time reflects the tension of the head dissections, yet also hints at their rewards:

Anatomy takes a nasty turn once we go above the neck. Not only does the information increase in detail like crazy (the skull is amazing in its intricacy—seemingly endless numbers of holes, indentations, seams, processes, all of which have beautiful Latin names and some kind of function for development, protection, collection), but the intensity of the dissections seems to multiply exponentially. Everything that I perceived to be difficult before pales, and by "difficult" I mean a host of things. Emotionally, the dissections are certainly far harder. We are peeling off the skin of the face, we are removing the scalp, we are tugging out the eyeball. But physically they are harder, too. We must saw around the circumference of the skull, chisel through the orbit, the jaw.

We chisel through the atlas, the perfectly named vertebra that holds the skull, and then pull the head up and forward so it flops down on the body's chest, now connected to the body only by the muscles that attach the sternum to the jaw, leaving a stump of neck one would imagine from a guillotine or an executioner's blade. Eventually we saw the head in half lengthwise, which is a gruesome and tiring task. I have some visceral reactions during these times, and so do my classmates.

The force necessary in the dissections feels barbarous, and I am still fascinated by what is revealed but hate the push and tug necessary for revelation. It feels wrong to push against the back of the head so strongly that you know the nose is smashed against the table; knowledge of pain's absence does nothing to remove the feeling that what you are doing is something that should not be done to a person, especially someone who has entrusted her body to you.

My dreams at night are full of Eve. I am Persephone, somewhere between the living and the dead. Feeling increasingly uneasy.

It is December, and the end of the term approaches. Mornings, a tight grip of frost clings to my windshield. It is still dark as I drive into campus, and my breath forms white clouds the moment it leaves my mouth—the way in which my body has changed the air made manifest. The snow on the ground looks blue in the darkness, except for the occasional yellow halos cast by the streetlights that send the shadow of my car, distorted and large, behind me as I drive. Even as the hours of darkness are lengthening, my nights of sleep have grown shorter and shorter. Exams are looming, and when I do turn out the last light in the house and, drained, climb into bed, even my dreams are overtaken by medicine. Odd mixtures of all of my coursework combine in fruitless nighttime anxiety, and when I wake, I am grouchy and unrestored—not even sleep is a respite from tension, from the feeling that some un-tended thing needs to be understood, or committed to memory.

The cranial bisection—the sawing of the skull into two sym-metrical halves—splits Eve's head between the eyes, divides the nose and lips and tongue and chin in two. When we reach the mouth and see perfectly pink gums, we realize for the first time that Eve still wore her dentures. Wordlessly, Trip pries them out with her fingers, sets them on the table beside Eve's right shoulder, and we continue the cleaving.

When the bisection of the head is complete, the few remain-ing days of the semester are spent in review for the final exam. My classmates and I have finished dissecting our cadavers. How-ever, the need to prepare for the upcoming exam and our fatigue from partitioning and studying the head combine to make this accomplishment anticlimactic. There is no celebration or com-memoration. We merely segue into reviewing the structures we have forgotten and committing to memory those we have recently exposed.

Initially I find my discomfort with the head dissections dis-couraging. After all, hadn't I been feeling increasingly at ease in the anatomy lab over the course of the semester? Now, as the course draws to a close, shouldn't I notice a real difference from those first, tentative days? I fear that my regression into fitful sleep and edginess somehow marks my failure as a doctor-in-training. Yet, increasingly, I am now able to hear the symptoms of a patient and have some sense of what might be wrong. I can pick up Eve's arm, or heart, and name its innermost structures; I can trace the flow of blood through her body, naming her arteries and veins. What does it mean that my emotional progress is more difficult to chart?

I think the reality of doctoring may be that in medical school, and in the years of training and practice that follow, clinical comfort steadily increases. Concepts that were initially elusive become clear; procedures that seemed impossible are eventually done with ease. Most importantly, a gestalt view of the body begins to build, so that injury and disease become easier to identify and the corresponding means of returning the body to health are more comprehensively understood. Nonetheless, comfort with certain wrenching tasks—the most difficult dissections, assessing a severely injured trauma patient, delivering a terminal diagnosis to a young person—must naturally wax and wane. I think that Deborah was right when she suggested that the emotional challenge of anatomy class might be partially intended to prepare my classmates and me for the rigors of tending to patients whose bodies are sick and maimed. Yet I also believe that the lesson of anatomy is that we do not need to overcome all our emotion or conquer all difficulty in order to be good clinicians. In fact, in light of the important balance that clinical detachment requires, I should perhaps feel encouraged by my inability to always emotionally disengage.

Eve's body has been thoroughly dissected, and we still do not know why she had no belly button. Dr. Goslow hypothesizes that she had some kind of abdominal surgery and that in the closing of the surgical wound the umbilicus got tucked in, like a seam. "But there's no scar," we protest.

He shrugs. "Maybe she had the surgery at such a young age that the scar just faded away."

We are not satisfied. There was no evidence of major surgery inside Eve's abdomen. There was no scar. And so it remains a mystery, a symbol of how some things about Eve remain unknowable, that our understanding of her cannot help but be only partial, even after the dissection is complete.

The semester ends in a blur of studying and exams, of sleep deprivation and note cards. When the final exam is behind us and I have officially finished my first semester of medical school, I find myself returning at unexpected moments to thoughts of Eve. And the thoughts are less and less the intrusive, troubling ones that interrupt my sleep and disrupt my mood, and more a deepening sense of gratitude and awe for Eve. Poor, dead Eve, who came to me beautiful and whom I have reduced to a shambles of muscle and bone, cut away and divided. Eve, who has, among other things, given me a sure and indisputable understanding of the impenetrable fact of death. I do not shudder at all to think of my body, or the bodies of my loved ones, burning to dust. Eve has shown me that no matter how gravely a dead body is altered, its lifelessness is the one aspect that does not budge.

I think of Eve's family in their mourning, some years ago, and cannot imagine that the thought of her body in the anatomy lab was to them either comforting or sacred. I wonder if her ability to donate her body came from the very knowledge that she gave to me: No harm can come to one after death. I wonder if she shared that knowledge with her family, and if it brought them peace.

I think of the vast amount of knowledge I've acquired since the year began. When my mother called me to tell me about the surgery on my grandfather's leg and said "femoral artery bypass," the vessel I pictured was not my grandfather's but Eve's. I have never seen the right middle cerebral artery that occluded to leave my grandmother's left side debilitated, but I have seen Eve's and can picture its precise path and the brain matter it nourished.

Truly, when I listen to any patient's heartbeat or lungs, or feel for someone's liver or pulse, or find tendons to tap with my hammer in order to test reflexes, the structures I picture hidden beneath the skin are all—all of them—Eve's. As Vesalius and William Harvey and Michelangelo and Leonardo da Vinci knew, the body's interior is a black box into which we grope and guess, unless we have had a chance to unveil it and see beyond the mystery. I cannot begin to know what led Eve to give me such a gift, whether it was practicality or altruism or cynicism or love of science or some other, equally unknowable, aspect of her personality or life. What I do know is that she neither knew me nor knew anything about me, and yet she bequeathed to me this offering, unthinkable for centuries, that has formed the foundations of my ability to heal. My hours with her neither cured her nor eased her suffering. Bit by bit, I cut apart and dismantled her, a beautiful old woman who came to me whole. The lessons her body taught me are of critical importance to my knowledge of medicine, but her selfless gesture of donation will be my lasting example of how much it is possible to give to a total stranger in the hopes of healing. That lesson, when I am called to treat critically ill patients who no longer appear human, and prisoners, and demented grandfathers who are dying and angry and scared, is the lesson I hope beyond all else to have absorbed.

Good-bye

And then there is using everything.

GERTRUDE STEIN

After the final anatomy exam, I punch the code into the doors of the lab and go into the room for the last time. I am alone. It is silent and dark. I stand for a minute in the doorway before turning on the lights. When I flip on the light switch, I see that the tables have all been moved into the center of the room.

I do not know which bag is Eve. I unzip one, then another, then another until I find her. When I do, I am not sure what I have come to do.

What part of myself was exposed in opening Eve? What structure in me was found and laid bare?

I touch her brain, resting in the upturned dome of her skull. I touch the bare bone of her face, her shoulder, her pelvis, her leg.

My hand comes to rest in hers. I feel her vessels beneath my fingers, her tendons and bones.

Great teacher, I give you flowers. I carry your body to the funeral pyre. When you burn, may every space in you that I have named flare and burst into light.

Bibliography

Abu-Hijleh, Marwan F., Nazih A. Hamdi, Satei T. Moqattash, Philip F. Harris, and Gilbert F. D. Heseltine. "Attitudes and Reactions of Arab Medical Students to the Dissecting Room." *Clinical Anatomy* 10:272–78, 1997.

Ariès, Philippe. *The Hour of Our Death*. Oxford: Oxford University Press, 1981.

Bidloo, Govard. *Godefridi Bidloo, medicinae doctoris & chirurgic, Anatomia humani corporis, centum & quinque tabuli*. Amsterdam, 1685.

The Burial at Thebes: Sophocles' Antigone translated by Seamus Heaney, Abbey Theatre Playscript Series. London: Faber and Faber, 2004.

Chaucer, Geoffrey. *The Canterbury Tales*, translated by Nevill Coghill. London: Penguin Books, 2003.

Cheney, Annie. "The Resurrection Men." *Harper's*, March, 2004.

Colapinto, John. *As Nature Made Him: The Boy Who Was Raised as a Girl*. New York: HarperCollins Publishers, 2000.

Cotran, Ramzi S., Vinay Kumar, and Tucker Collins. *Robbins Pathologic Basis of Disease*. W. B. Saunders, 1999.

Dibdin, Michael. "The Pathology Lesson." *Granta* 39 (Spring 1992) 99.

Dickinson, George E., Carol J. Lancaster, Idee C. Winfield, Eleanor F. Reece, and Christopher A. Colthorpe. "Detached Concern and Death Anxiety of First-Year Medical Students: Before and After the Gross Anatomy Course." *Clinical Anatomy* 10:201–7, 1997.

Doty, Mark. *Atlantis: Poems*. New York: HarperCollins, 1995.

——— *Heaven's Coast*. New York: Harper Perennial, 1997.

Dreger, Alice Domurat. *Hermaphrodites and the Medical Invention of Sex*. Cambridge, Mass.: Harvard University Press, 1998.

Druce, Maralyn, and Martin H. Johnson. "Human Dissection and Attitudes of Preclinical Students to Death and Bereavement." *Clinical Anatomy* 7:42–49, 1994.

Dyer, George S. M., and Mary E. L. Thorndike. "Quidne Mortui Vivos Docent? The Evolving Purpose of Human Dissection in Medical Education." *Academic Medicine* 75 no. 10 (October 2000): 969–79.

Evans, E. J., and G. H. Fitzgibbon. "The Dissecting Room: Reactions of First Year Medical Students." *Clinical Anatomy* 5:311–20, 1992.

Ferrari, Giovanna. "Public Anatomy Lessons and the Carnival: The Anatomy Theatre of Bologna." *Past and Present* no. 117 (November 1987): 50–106.

Finkelstein, Peter, and Lawrence Mathers. "Post-Traumatic Stress Among Medical Students in the Anatomy Dissection Laboratory." *Clinical Anatomy*, 3:219–26, 1990.

Gatrell, V. A. C. *The Hanging Tree: Execution and the English People, 1770–1868*. Oxford: Oxford University Press, 1994.

Gogarty, Oliver St. John. *Tumbling in the Hay*. 1939, reprinted by Sphere Books, 1982.

Gonzalez-Crussi, F. *Notes of an Anatomist*. London: Picador, 1986.

Hansen, John T. *Essential Anatomy Dissection: Following Grant's Method*. Baltimore: Williams and Wilkins, 1998.

Hyman, Libbie Henrietta. *Comparative Vertebrate Anatomy*. Chicago: University of Chicago Press, 1922.

Kemp, Martin, and Marina Wallace. *Spectacular Bodies*. Hayward Gallery, London, and Berkeley: University of California Press, 2000.

Larson, Erik. *Devil in the White City*. New York: Vintage Books, 2003.

Lynch, Thomas. *The Undertaking: Life Studies from the Dismal Trade*. New York: Penguin, 1998.

Marks Jr., Sandy C., Sandra L. Bertman, and June C. Penney. "Human Anatomy: A Foundation for Education About Death and Dying in Medicine." *Clinical Anatomy* 10:118–22, 1997.

McGarvey, M. A., T. Farrell, R. M. Conroy, S. Kandiah, and W. S. Monkhouse. "Dissection: A Positive Experience." *Clinical Anatomy* 14:227–30, 2001.

Miller, Jonathan. *The Body in Question*. New York: Random House, 1978.

Nhat Hanh, Thich. *The Miracle of Mindfulness*. Boston: Beacon Press, 1975.

Nnodim, J. O., "Preclinical Student Reactions to Dissection, Death, and Dying." *Clinical Anatomy* 9:175–82, 1996.

Nuland, Sherwin. *Doctors: The Biography of Medicine*. New York: Vintage Books, 1988.

O'Carroll, R. E., S. Whiten, D. Jackson, and D. W. Sinclair. "Assessing the Emotional Impact of Cadaver Dissection on Medical Students." *Medical Education* 36:550–54, 2002.

Petherbridge, Deanna, and Ludmilla Jordanova. *The Quick and the Dead: Artists and Anatomy*, Hayward Gallery, London, and Berkeley: University of California Press, 1997.

Pickering, David. *Cassell Dictionary of Superstitions*. London: Cassell Wellington House, 1995.

Richardson, Ruth. *Death, Dissection and the Destitute*. Chicago: University of Chicago Press, 2000.

Rizzolo, Lawrence J. "Human Dissection: An Approach to Interweaving the Traditional and Humanistic Goals of Medical Education." *Anatomical Record* 269:242–48, 2002.

Rohen, Johannes W., Chihiro Yokochi, and Elke Lütjen-Drecoll. *Color Atlas of Anatomy, A Photographic Study of the Human Body*. Philadelphia: Lippincott Williams & Wilkins, 1998.

Rufus, Anneli. *Magnificent Corpses*. New York: Marlowe, 1999.

Sappol, Michael. *A Traffic of Dead Bodies: Anatomy and Embodied Social Identity in Nineteenth-Century America*. Princeton: Princeton University Press, 2002.

Sawday, Jonathan. *The Body Emblazoned: Dissection and the Human Body in Renaissance Culture*. London: Routledge, 1995.

Snell, Richard S. *Clinical Anatomy for Medical Students*, 6th ed. Philadelphia: Lippincott Williams & Wilkins, 2000.

Stone, Irving. *The Agony and the Ecstasy*. Garden City, NY: Doubleday, 1961.

Vesalius, Andreas. *De humani corporis fabrica*. Basileae: Ex officina Ioannis Oporini, 1543.

———, William Frank Richardson, and John Burd Carman. *On the Fabric of the Human Body, Book I: The Bones and Cartilages*. San Francisco: Jeremy Norman, 1998.

———. *On the Fabric of the Human Body, Book II: The Ligaments and Muscles*. San Francisco: Jeremy Norman, 1999.

Widdess, J. D. H. *The Royal College of Surgeons in Ireland and its Medical School 1784–1984*, 3rd ed. Dublin: Royal College of Surgeons Publications Department, 1984.

Wilf, Steven Robert. "Anatomy and Punishment in Late Eighteenth-Century New York." *Journal of Social History* 22 (Spring 1939): 507–31.

Winkelmann, Andreas, and Fritz H. Güldner. "Cadavers as Teachers: The Dissecting Room Experience in Thailand." *British Medical Journal* 329: 1455–57, 2004.

Winterson, Jeanette. "The Cells, Tissues, Systems and Cavities of the Body." *Granta* 39 (Spring 1992).

Wise, Sarah. *The Italian Boy: A Tale of Murder and Body-Snatching in 1830s London*. New York: Metropolitan Books, 2004.

Yardley, Jim. "Dead Bachelors in Remote China Still Find Wives." *New York Times*, October 5, 2006, page A1.

Acknowledgments

Attending medical school and writing a book are both daunting undertakings; I consider myself matchlessly fortunate to have met and befriended Ted Goslow, who championed my efforts at both. His expertise and endless enthusiasm for me and for my project will be difficult gifts to repay.

I owe sincere thanks to Brown University and the department of ecology and evolutionary biology. In particular I would like to thank Dr. Edward Feller, Dr. Melvin Hershkowitz, Dale Ritter, Alex Morang, Tripler Pell, Lexy Westphal, Shannon Silva, Dean Marsh, Jennifer Rankin, Amygdala Interactions, and the Brown Summer Fellowship for the Arts. The illustrations in the book appear courtesy of the John Hay Library at Brown, where Andy Moul and Richard Noble helped me gain access to the remarkable Lownes Collection of rare books.

In writing about the history of dissection, and in particular the resurrection trade, I acknowledge my heavy debt to the fascinating and masterfully written study by Ruth Richardson, *Death, Dissection and the Destitute* (University of Chicago Press, 2000).

Professor Clive Lee, Alice McGarvey, and the histopathology and anatomy departments at the Royal College of Surgeons in Ireland kindly shared their resources with me, from bodies to books. In addition, my European research was facilitated by Anna Lerose, Cora Castaldi, Stephan Blatti, and Cindy Klestinec.

The business of books is a foreign land to me, and I consider myself especially lucky to have been shepherded through it by my wonderful agent, Kris Dahl. Emily Loose and Janie Fleming both provided invaluable editorial insight to my manuscript as it took shape. And I owe special thanks for guidance and encouragment to Mark Shaw and the Books for Life Foundation, Winifred Gallagher, and the Lafage Writer's Retreat.

My writing and my life have benefited from the unquantifiable support of Scott, Janice, Eric, Laura, Andrew, Sarah, and Bill Montross, and the indomitable Mary Townsend.

Finally, my deepest gratitude and love to Deborah, who endured sleep disturbances; clothes that smelled of formalin; time in lovely Italian cities spent in anatomy theaters, creepy wax museums, and in constant quest for stranger and more gruesome reliquaries; and, most of all, a personal life that consisted quite disproportionately of discussions of dead bodies and their various fates. Her vision for and confidence in this book—as in our lives—were often greater than my own.

And to Maude, who is simply a dream come true.